Lecture Notes in Control and Information Sciences

Edited by A. V. Balakrishnan and M. Thoma

Vol. 22: Optimization Techniques
Proceedings of the 9th IFIP Conference on Optimization Techniques,
Warsaw, September 4–8, 1979
Part 1
Edited by K. Iracki, K. Malanowski, S. Walukiewicz
XVI, 569 pages. 1980

Vol. 23: Optimization Techniques
Proceedings of the 9th IFIP Conference on Optimization Techniques,
Warsaw, September 4-8, 1979
Part 2
Edited by K. Iracki, K. Malanowski, S. Walukiewicz
XV, 621 pages. 1980

Vol. 24: Methods and Applications in Adaptive Control
Proceedings of an International Symposium
Bochum, 1980
Edited by H. Unbehauen
VI, 309 pages. 1980

Vol. 25: Stochastic Differential Systems – Filtering and Control
Proceedings of the IFIP-WG7/1 Working Conference
Vilnius, Lithuania, USSR, Aug. 28 – Sept. 2, 1978
Edited by B. Grigelionis
X, 362 pages. 1980

Vol. 26: D. L. Iglehart, G. S. Shedler
Regenerative Simulation of Response Times in Networks of Queues
XII, 204 pages. 1980

Vol. 27: D. H. Jacobson, D. H. Martin, M. Pachter, T. Geveci
Extensions of Linear-Quadratic Control Theory
XI, 288 pages. 1980

Vol. 28: Analysis and Optimization of Systems
Proceedings of the Fourth International Conference on Analysis and Optimization of Systems
Versailles, December 16–19, 1980
Edited by A. Bensoussan and J. L. Lions
XIV, 999 pages. 1980

Vol. 29: M. Vidyasagar,
Input-Output Analysis of Large-Scale Interconnected Systems – Decomposition, Well-Posedness and Stability
VI, 221 pages. 1981

Vol. 30: Optimization and Optimal Control
Proceedings of a Conference Held at Oberwolfach, March 16–22, 1980
Edited by A. Auslender, W. Oettli, and J. Stoer
VIII, 254 pages. 1981

Vol. 31: Berc Rustem
Projection Methods in Constrained Optimisation and Applications to Optimal Policy Decisions
XV, 315 pages. 1981

Vol. 32: Tsuyoshi Matsuo,
Realization Theory of Continuous-Time Dynamical Systems
VI, 329 pages, 1981

Vol. 33: Peter Dransfield
Hydraulic Control Systems – Design and Analysis of Their Dynamics
VII, 227 pages, 1981

Vol. 34: H.W. Knobloch
Higher Order Necessary Conditions in Optimal Control Theory
V, 173 pages, 1981

Vol. 35: Global Modelling
Proceedings of the IFIP-WG 7/1 Working Conference Dubrovnik, Yugoslavia, Sept. 1–5, 1980
Edited by S. Krčevinac
VIII, 232 pages, 1981

Vol. 36: Stochastic Differential Systems
Proceedings of the 3rd IFIP-WG 7/1 Working Conference
Visegrád, Hungary, Sept. 15–20, 1980
Edited by M. Arató, D. Vermes, A.V. Balakrishnan
VI, 238 pages, 1981

Vol. 37: Rüdiger Schmidt
Advances in Nonlinear Parameter Optimization
VI, 159 pages, 1982

Vol. 38: System Modeling and Optimization
Proceedings of the 10th IFIP Conference
New York City, USA, Aug. 31 – Sept. 4, 1981
Edited by R.F. Drenick and F. Kozin
XI, 894 pages. 1982

Vol. 39: Feedback Control of Linear and Nonlinear Systems
Proceedings of the Joint Workshop on Feedback and Synthesis of Linear and Nonlinear Systems
Bielefeld/Rom
XIII, 284 pages. 1982

Vol. 40: Y.S. Hung, A.G.J. MacFarlane
Multivariable Feedback: A Quasi-Classical Approach
X, 182 pages. 1982

Vol. 41: M. Gössel
Nonlinear Time-Discrete Systems – A General Approach by Nonlinear Superposition
VIII, 112 pages. 1982

Vol. 42: Advances in Filtering and Optimal Stochastic Control
Proceedings of the IFIP-WG 7/1 Working Conference
Cocoyoc, Mexico, February 1–6, 1982
VIII, 391 pages. 1982

For information about Vols. 1–21 please contact your bookseller or Springer-Verlag.

Lecture Notes in Control and Information Sciences

Edited by M. Thoma

73

J. Zarzycki

Nonlinear Prediction Ladder-Filters for Higher-Order Stochastic Sequences

Springer-Verlag
Berlin Heidelberg GmbH

Author
Jan Zarzycki
Institute of Telecommunication and Acoustics
The Technical University of Wroclaw
ul. B. Prusa 53/55
50-317 Wroclaw – Poland

ISBN 978-3-540-15635-2 ISBN 978-3-540-39500-3 (eBook)
DOI 10.1007/978-3-540-39500-3

Library of Congress Cataloging in Publication Data

Zarzycki, J. (Jan)
Nonlinear prediction ladder-filters for higher-order stochastic sequences.
(Lecture notes in control and information sciences; 73)
Bibliography: p.
1. Stochastic sequences. 2. Prediction theory.
3. Filters (Mathematics) I. Title. II. Series.
QA274.225.Z37 1985 519.2 85-12668

Originally published by Springer-Verlag Berlin Heidelberg New York in 1985.

2161/3020-543210

PREFACE

In this work we shall be concerned with the problem of nonlinear least-squares prediction of higher-order stochastic sequences using nonlinear orthogonal digital filters. The nonlinear problem will be considered as a generalization of the linear least-squares prediction problem.

The linear least-squares estimation theory is associated with second-order stochastic sequences, for which the orthogonal prediction or innovations linear filters (producing white noise when driven by the given sequence) as well as modeling or shaping filters (whose output is statistically equivalent to the given sequence, when driven by white noise) can be established.

In practice, the linear orthogonal filters can be computed via recursive procedures, in which one will not have to recompute the whole filter each time the permitted complexity is increased. Hence, the same idea underlies both the ladder-structures implemented in modern digital filters theory, and the theory of orthogonal (Fourier) expansions in Hilbert spaces. One of the most important property of the orthogonal digital filter is preservation of 'energy' which assures inherent numerical stability of the filter. A remarkable result of this theory is that any transfer function can be realized by means of an orthogonal filter whose modular structure can be implemented using sophisticated 'building-blocks' with VLSI integrated circuits (namely CORDICS processors).

The linear theory results in the optimum (least-squares) stochastic approximation of second-order sequences. Therefore, the linear estimation filter becomes the best possible filter for a Gaussian sequence (whose properties are completely characterized by the second-order statistics). If the underlying sequence is non-Gaussian, the linear estimation accuracy may be not satisfactory. In that case, a nonlinear approach to the problem should be introduced in order to improve the accuracy.

In this work we wish to present efficient algorithms of nonlinear least-squares prediction filters for higher-order stochastic sequences, resulting in the optimum approximate nonlinear digital filters of the Volterra-Wiener class. These nonlinear ladder-filters will generalize the linear filters, preserving most of their properties (orthogonality and modular realizations, among others), and yielding better estimation accuracy for higher-order (and non-Gaussian) stochastic sequences.

We will mention here only those papers which are closely connected to the subject of this work, reffering for more complete bibliography to the papers citied (and the references therein).

ACKNOWLEDGMENTS

I am particularly indebted to Professor Patrick Dewilde of the Delft University of Technology for his helpful suggestions and hints introduced in many stimulating and fruitful discussions, especially during my one-year stay in Delft, which has undoubtedly inspired this work.

I am also grateful to Professor Marian S. Piekarski of the Technical University of Wroclaw for his valuable comments and discussions concerning this work.

I wish to thank Mrs. Zdzislawa Żabska for her careful typing of this manuscript.

CONTENTS

1. INTRODUCTION

Let $(\Omega, \mathcal{B}, \mu)$ denote a probability space where Ω in an abstract set whose elements are denoted by υ, $\mathcal{B}$ is a σ-algebra of subsets of Ω, and μ is a probability measure on $\mathcal{B}$. By $\mathbb{L}_2(\Omega, \mathcal{B}, \mu)$ we will understand a collection of σ-measurable maps $w: \Omega \to \mathbb{R}$ for which $\int_\Omega |w(\upsilon)|^2 d\mu < \infty$. We introduce inner-product on $\mathbb{L}_2(\Omega, \mathcal{B}, \mu)$ as $(w,v) \triangleq \int_\Omega w(\upsilon)\bar{v}(\upsilon) d\mu = \mathbb{E}\{w\bar{v}\}$, bar standing for complex conjugation. This inner-product will induce the norm $\|w\|^2 = \mathbb{E}\{|w|^2\}$, and the metric $d(w,v) = \|w-v\|$. Assuming completeness, $\mathbb{L}_2(\Omega, \mathcal{B}, \mu)$ will be a Hilbert space. Let $t \in T$, T being some index-set. A map $t \to y_t = w_t(\upsilon) \in \mathbb{L}_2(\Omega, \mathcal{B}, \mu)$ will be called a Hilbert stochastic sequence if for all $t \in T$, $\mathbb{E}\{|y_t|^2\} < \infty$. That sequence will be denoted by $\{y\}$. The collection $\{y\}$ will be called a K-th order stochastic sequence if its properties will be described by the joint probability distributions of the subfamilies $\{y_{t_1}, \dots, y_{t_k}\}$, $k=1,\dots,K$ so that the first K averages $\mathbb{E}\{y_{t_1} \dots y_{t_k}\}$, $k=1,\dots,K$ are assumed to be known. Two stochastic sequences $\{y\}$ and $\{x\}$ will be called statistically equivalent in a weak K-th order sense if for $k=1,\dots,K$ we will have the following equalities $\mathbb{E}\{y_{t_1} \dots y_{t_k}\} = \mathbb{E}\{x_{t_1} \dots x_{t_k}\}$.

The problem of prediction is to compute $\hat{y}_t = y_{t|t-1,t-2,\dots}$, a random variable being a fixed function from $\{y\}$ conditional upon some 'past' subfamily $\{y_{t-1}, y_{t-2}, \dots\}$. Stating the problem geometrically, within the Hilbert space framework, the estimate will be optimal if $\hat{y}_t$

will be the orthogonal projection upon the closed subspace spanned by that 'past'. In that case, the length $\|e_t\|^2 = \mathbb{E}\{|e_t|^2\}$ of the error $e_t \triangleq y_t - \hat{y}_t$ will be minimized. Denoting by $\mathbf{Y}$ the subspace spanned by $\{y_{t-1}, y_{t-2}, \ldots\}$, and by P the orthogonal projection operator taking projection on the subspace $\mathbf{Y}$, the optimum estimate of y_t will be $\hat{y}_t = P\, y_t$ since the coprojection $e_t = P^{\perp} y_t$ will be orthogonal to $\mathbf{Y}$. Consequently, the problem of prediction is to compute that projection.

Projection of y_t on $\mathbf{Y}$ determines a prediction filter, operating on the 'past' of the sequence $\{y\}$, and yielding the 'best' (in the least-squares sense) approximation of that random variable. The notion of 'best' approximation relies naturally on a notion of 'distance' between the 'ideal' prediction filter F (whose output is just y_t), and the optimum approximate prediction filter F^a (producing $\hat{y}_t$). The distance between the two filters can be asserted to be the least-squares discrepancy between their outputs, averaged over the entire input collection; i.e., $\mathbb{E}\{|y_t - \hat{y}_t|^2\}$. This observation permits one to convert the problem of 'best' stochastic approximation (of the random variable y_t) into the problem of 'best' deterministic approximation (of the 'ideal' filter F) in a space of filters.

Filters, considered as devices mapping the space of excitations into the space of responses form a vector space in a natural way, see e.g., Victor and Knight (1979). Addition in that space is equivalent to parallel connection of filters, a scalar multiplication corresponds to a change of the filter gain. The choice of input sequence (i.e., of the probability space from which the excitations are taken) yields naturally a notion of 'distance' between the filters $\|F_1 - F_2\|^2 = \mathbb{E}\{|F_1(\cdot) - F_2(\cdot)|^2\}$ where $\mathbb{E}$ indicates the average over the input probability space. The inner-product on the space $\mathbb{F}$ of filters will be introduced as follows $(F_1,F_2) \triangleq \mathbb{E}\{F_1(\cdot)\bar{F}_2(\cdot)\}$, and will induce the norm $\|F\|^2 = \mathbb{E}\{|F|^2\}$. We notice that we might have $(F,F) = 0$ for some non-zero filters, however, their outputs will be a.e. zero in the ensemble of excitations. Such filters will form a subspace $\mathbb{F}_o$. That subspace will contain all filters which are not distinguishable from the zero-filter in an identification scheme implied by the input ensemble. Consequently, we will consider the quotient space $\mathbb{F}' = \mathbb{F}/\mathbb{F}_o$ of the distinguishable filters (relative to the ensemble of the underlying excitations). The space $\mathbb{F}'$ may now be completed to form a Hilbert space. Hence, the stochastic approximation problem in the space of the random variables can now be treated as the filter approximation problem in the space of filters.

Considering the elements of the space of filters $\mathbb{F}'$ as operators or, equivalently, as time-indexed functionals, it follows from Frechet (1910) that the regular Volterra functional polynomials

$$G_{M;t} = \sum_{m=1}^{M} V_{m;t} \tag{1.1}$$

where

$$V_{m;t} = \sum_{j_1} \cdots \sum_{j_m} a_{m;j_1 \dots j_m} \, y_{j_1} \cdots y_{j_m} \tag{1.2}$$

with $a_{m;j_1 \dots j_m} = 0$ for $j_r > t$, $r=1,\dots,m$ (see Volterra (1959)) span a dense subspace in the space of filters. Hence, that space is separable, and the Volterra functionals will form a countable complete set (basis). This is a natural generalization (to functionals) of the Weierstrass theorem, stating that polynomials span a dense subspace in the space of (sufficiently regular) functions so that the space is separable, and polynomials form a complete countable set.

Consequently, each filter F (for which such representation exists) can be approximated arbitrarily well by the Volterra series

$$F = \sum_{m=1}^{\infty} V_m \qquad (1.3)$$

This means that the given filter F is approximated in the subsequent subspaces $\vee\{V_m\}$ whose elements are nonlinear Volterra-type filters of the subsequent degrees of nonlinearity. We notice that $\vee\{V_1\}$ will be the subspace of linear filters; $\vee\{V_2\}$ will consists of the second-degree nonlinear filters, etc. Computation of the subsequent terms in the series (1.3) will therefore consist in determination of the set of multi-dimensional (M-D) impulse responses, see e.g., Schetzen (1980),

$$\{ a_{t;j_1 \ldots j_m} , \; m=1,2,\ldots \} \qquad (1.4a)$$

of the approximate nonlinear filter F^a (in time-domain) or, equivalently, of the set of M-D transfer functions

$$\{ A_t(e^{i\theta_1},\ldots,e^{i\theta_m}) , \; m=1,2,\ldots \} \qquad (1.4b)$$

(in frequency-domain), in the time-variant (nonstationary) case. If the input sequence is stationary, the approximate nonlinear filter will be time-invariant, and will be represented by the sets (1.4) provided the variable t is removed (following the shift-invariance of inner-product in stationary case). We remark that approximation of the 'ideal' filter F by means of a linear filter will restrict the considerations to the subspace $\vee\{V_1\}$ only. In that case, the approximate linear filter will be described by the 1-D members of the sets (1.4).

The series-expansion (1.3) is valid only for the analytic (in the sense of Frechet or Gateaux) elements F of the space of filters $\mathbb{F}$! For non-analytic filters, the Wiener-type expansion should be considered. This sort of expansion can be introduced via (Gram-Schmidt) orthogonalization of the Volterra functional polynomials G_M (1.1) relative to the inner-product implied by the input probability space. Orthogonalization of the Volterra-type basis will yield a countable complete orthonormal set $\{W_M, M=1,2,...\}$, being actually an ON basis of the space of filters, and implying the orthogonal decomposition

$$\mathbb{F}' = \sum_{M=1}^{\infty} \oplus \vee \{W_M\} \tag{1.5}$$

where $(W_M, W_N) = \delta_{M,N}$, $\delta_{M,N}$ being Kronecker delta. Each member W_M of that ON basis will span the orthogonal subspace with the elements being the M-th degree nonlinear orthogonal filters of the Volterra-Wiener class. Hence, any filter F, for which $\| F \|^2 < \infty$, will be represented by the Wiener-type orthogonal series-expansion

$$F = \sum_{M=1}^{\infty} (F, W_M)\, W_M \tag{1.6}$$

which is actually the stochastic functional Fourier series, with (F, W_M) being the generalized M-D Fourier kernel. In other words, the 'ideal' filter F will be represented in terms of orthogonal series of nonlinear orthogonal filters (for a given input sequence). That series is convergent for a wider class of filters (functionals) than the class of analytic filters, much like the 'usual' Fourier series is convergent for a wider class of functions than the functions expressible in terms of convergent Taylor series. The orthogonal expansion (1.6) can be partitioned into the two, mutually orthogonal, components (relative to the M-th term)

$$F = \sum_{m=1}^{M} (F, W_m)\, W_m + \sum_{m=M+1}^{\infty} (F, W_m)\, W_m \tag{1.7}$$

The first component will express the optimum M-th degree nonlinear orthogonal estimate F_M^a of the 'ideal' filter F, and can be interpreted as the orthogonal projection of F on the subspace spanned by the subset $\{W_1,\dots,W_M\}$. The second component will express the approximation error. In other words, the first RHS term in (1.7) will be the optimal M-th degree nonlinear approximate filter; i.e.,

$$F \sim F_M^a = \sum_{m=1}^{M} (F,W_m)\, W_m \tag{1.8}$$

Observing that the output of the 'ideal' filter F equals y_t while the optimum M-th degree nonlinear estimate $\hat{y}_{M;t}$ is the output of the approximate filter F_M^a, we notice that computation of that approximate nonlinear filter will yield the optimum M-th degree nonlinear solution to the stochastic approximation problem since the norm of the error $e_{M;t}$ (corresponding to the second RHS term in (1.7)) will be minimized. Hence, computation of (1.8) solves the problem of the optimum prediction of the 2M-th order stochastic sequence, provided the first 2M covariances of that sequence (i.e., $\mathbb{E}\{y_{j_1}\dots y_{j_m}\bar{y}_{k_1}\dots\bar{y}_{k_u}\}$, $m,u=1,\dots,M$) are given.

It should be noted that the linear least-squares prediction problem will correspond to the optimum approximation of the 'ideal' prediction filter F by means of the linear approximate prediction filter F_1^a. In that case the considerations will be restricted to the subspace $v\{W_1\}$ only, so that the linear problem will correspond to $M=1$ in (1.7), and

$$F \sim F_1^a = (F,W_1)\, W_1 \tag{1.9}$$

Consequently, the linear estimation problem will be associated with second-order sequences with the covariance data $\mathbb{E}\{y_{j_1}\bar{y}_{k_1}\}$. If the second-order statistics are sufficient in order to characterize the underlying sequence completely (the Gaussian case), the linear approximate filter F_1^a will become the best possible filter. In case of higher-order and non-Gaussian sequences, the linear estimation accuracy may be not satisfactory. In that case, estimation accuracy may be improved by replacing the linear estimation scheme by the nonlinear procedure, introducing higher-order terms corresponding to M=2,3,... . This follows from the fact that in the linear case, the norm of the error is reduced by the first Fourier kernel only; i.e.,

$$\|e_{1;t}\|^2 = \|y_t\|^2 - |(F,W_1)|^2 \tag{1.10a}$$

while in the M-th degree nonlinear case we will have

$$\|e_{M;t}\|^2 = \|y_t\|^2 - |(F,W_1)|^2 - \dots - |(F,W_M)|^2 \tag{1.10b}$$

Orthogonal representation of a $\mathbb{L}_2$-functional is due to Wiener (1942,1958) who introduced Gram-Schmidt orthogonalization process for the Volterra functional polynomials of the Wiener process. The work of Wiener was preceded by Cameron and Martin (1947) who proposed the orthogonal expansion of the $\mathbb{L}_2$-functional using CON set of multi-variate Hermite polynomials, following Kaczmarz and Steinhaus (1935), and Wiener (1938). The problem was considered from a deeper mathematical viewpoint by Ito (1951) who introduced direct orthogonal representation, and showed its relation to the Fourier-Hermite series of Cameron and Martin. Introducing the Volterra functional polynomials into the latter series, Wiener (1958) presented his theory of the orthogonal approximation of a $\mathbb{L}_2$-functional in the Wiener-series form, underlying the theory of the orthogonal nonlinear systems with memory. Barret (1963) proposed the ON development for a functional of a stationary sequence, using the set of CON polynomial-functionals. The orthogonal 'gate-functions' approach was introduced by Bose (1958). Ogura (1972) considered the ON expansion for a functional of Poisson process using Charlier polynomials. The orthogonal development for a functional of independent-increment processes was introduced by Segall and Kailath (1976). Stochastic functional Fourier series for a functional of 'orthogonalizable' processes was discussed by Yasui (1979). Orthogonal expansions of that sort were also considered by Hida and Ikeda (1965), and by Victor and Knight (1979). The Volterra and Wiener theories have been investigated and developed in many works, yielding the theory of the nonlinear systems with memory, later termed the nonlinear systems of the Volterra-Wiener class (see Schetzen (1980), and the references therein).

In this work we wish to introduce efficient algorithms for computation of the nonlinear orthogonal approximate prediction filters (1.8) of the Volterra-Wiener class, for higher-order stochastic sequences observed on a finite time-interval. We will derive recursive algorithms of the nonlinear digital ladder-filters, in which estimation accuracy can be improved succesively via updating the order of the filter. We will consider nonlinear prediction filters with orthogonal structures, where only 'new' sections are introduced to the filter each time the order of the filter is updated, without altering the existing structure and its parameters. Consequently, we wish to present a class of nonlinear orthogonal prediction filters, yielding better (than in the linear treatment) estimation accuracy for higher-order and non-Gaussian sequences on one hand, on the other - obeying the attractive properties of modern linear orthogonal digital filters.

The orthogonal linear ladder-filter algorithms can be derived in time and/or in frequency-domains, using algebraic and/or geometric methods. The Levinson (1947) algorithm can serve as an example of time-domain algebraic solution in the autoregressive (AR) stationary case, see also Kailath (1974). If the underlying sequence is α-stationary (see Kailath (1982)), the Schur parametrization method (see Lev-Ari (1982); Kailath (1982); Lev-Ari and Kailath (1982)) can be applied in order to obtain efficient realizations of the linear innovations/modeling ladder-filters. The generalized Schur/Levinson methods of Deprettere and Lie (1980); Delsarte, Genin and Kamp (1983) can be used in nonstationary case. The Levinson method is closely connected to the theory of ON Szegő polynomials, and related to the Schur (1917) algorithm (see Dewilde, Vieira and Kailath (1978); Delsarte, Genin and Kamp (1979a)). The linear filter problem can be equivalently considered in the space of functions square-integrable w.r. to the spectral measure of the underlying second-order stationary stochastic sequence, under the Kolmogorov isomorphism. Consequently, the linear stochastic estimation problem can be converted into the deterministic approximation problem, yielding the cascade realizations of the linear filters (see Deprettere and Dewilde (1979), Dewilde and Dym (1981); Deprettere (1981); Dewilde (1982)). The problems of Darlington and cascade synthesis, and interpolation with positive real matrices, considered in Piekarski (1971,1974); Piekarski and Saeed (1980; Piekarski and Uruski (1984), are closely connected to those topics. The Levinson-Szegő theory, resulting in the approximate realizations of the AR and moving average (MA) filters and equivalent to the Schur method, can be generalized to the rational (ARMA) case (see Dewilde, Vieira and Kailath (1978); Dewilde and Dym (1981)). This approach, introducing the Schur algorithm in geometric context, underlies the theory of the linear ARMA innovations/modeling orthogonal filters, which can be applied in order to solve the lossless inverse scattering problem; the Nevanlinna-Pick interpolation problem; the optimum stochastic modeling problem; the maximum entropy spectral approximation problem, in discrete- and/or in continuous-time cases (see Dewilde and Dym (1981,1984), Dewilde (1982,1984); Deprettere (1981); Morf, Vieira, Lee and Kailath (1978); Kailath (1982);

Delsarte, Genin and Kamp (1979b); Deprettere and Dewilde (1979), Bultheel and Dewilde (1979); Prabhakara Rao and Helmond (1983); Widya (1982), and the references therein). The linear ladder-filter algorithms can be equivalently derived using projection method in time/or in frequency-domains (see Delosme and Morf (1982); Kailath (1982); Dewilde (1982)). The adaptive properties of the filters can also be considered (see Lee, Morf and Friedlander (1981); Kailath (1982); Deprettere and Lie (1980); Deprettere (1982)). The orthogonal linear ladder-filters can be efficiently implemented using VLSI technology, if the standard filter sections are interpreted in terms of hyperbolic or circular rotations, and then realized using CORDICS processors (see Wolder (1959); Tuszynski (1980); Ahmed and Morf (1982); Lev-Ari (1983); Dewilde (1982,1983,1984b); Deprettere (1983); Deprettere, Dewilde and Udo (1984); Deprettere and Jainandunsing (1984)).

Generalizing the linear ladder-filter algorithms, the nonlinear prediction filter problem can be stated and solved algebraically (see Zarzycki and Dewilde (1983); Zarzycki (1983)) or, equivalently, using projection method (Zarzycki (1984a-e,1985a)) yielding orthogonal representations of the nonlinear prediction filters of the Volterra-Wiener class (see Zarzycki (1985b)).

This work can be outlined as follows:

In chapter 2 we introduce the nonlinear least-squares prediction problem algebraically, and geometrically in the three following isomorphic vector spaces: of the regular Volterra functional polynomials; of generalized (block, multi-indexed) matrices; and of generalized (block, multi-variate) z-polynomials. We show that the nonlinear stochastic estimation problem in the first space can be equivalently considered as a deterministic problem of optimum approximation of the set of M-D impulse responses of the filter in the second space. In the latter space, this will become a deterministic problem of polynomial approximation of the set of M-D transfer functions of that filter.

In chapter 3 we present recursive geometric solutions to the nonstationary nonlinear prediction problem, using projection method in the spaces of the generalized matrices, and of the generalized z-polynomials. We derive a nonstationary nonlinear ladder-filter algorithm, yielding the nonlinear (Levinson) time-variant prediction filter with the orthogonal structure.

Chapter 4 is devoted to the problem of complexity reduction in the nonlinear ladder-filter. We introduce a time-variant nonlinear filter algorithm, following the shift-invariance of inner-products in the higher-order stationary case. Further complexity reduction is achieved by introducing a class of 'quasi-linear' ladder-filters, associated with the optimum prediction of higher-order stochastic sequences whose 'distance' from the Gaussian sequence is low (in a sense to be defined).

In Appendix 1 the generalized (multi-indexed) matrix theory is outlined while in Appendix 2 the 'local' order-update recursions and corresponding nonlinear filter sections are presented.

USAGE:

Capitals $\mathbb{Y},\mathbb{Z},\ldots$ will usually denote spaces; capitals $A,H,\ldots$ will indicate multi-indexed (multi-dimensional) matrices while lower case letters will be used for their entries. ${}^{m}A$ will denote a m-indexed matrix with entries $a_{j_1 \ldots j_m}$, and will be interpreted as a map ${}^{m}A : D^{m}A \to \mathbb{R}$, mapping the m-variate index-set $D^{m}A$ into the reals $\mathbb{R}$. The $D^{m}A$ will therefore be termed a 'domain' of the ${}^{m}A$. Thus, ${}^{1}A$ will be a row-matrix (or vector); ${}^{2}A$ will denote a rectangular (or 'usual') matrix; ${}^{3}A$ will stand for a cube matrix; etc. Then, ${}^{\{M\}}A = [{}^{1}A \ \ldots \ {}^{M}A]$ will be a generalized block (row) matrix whose block-entries are m-indexed matrices. ${}^{\{M\times M\}}A$ will stand for a generalized block (square) matrix with multi-indexed block-entries. $\mathbb{T}$ will indicate the unit-circle $\mathbb{T} = \{z : |z| = 1\}$ while ${}^{m}\mathbb{T}$

will be the unit m-torus $^{m}\mathbb{T} = \mathbb{T} \otimes \dots \otimes \mathbb{T}$. Unless otherwise noted all integrals are taken over the unit-circle $\mathbb{T}$ and, hence, the limits of integration are not indicated. $\{y\}$ will denote a stochastic sequence; i.e., a collection of random variables y_j $j \in J$ (J being some index-set). All stochastic sequences are assumed to be purely stochastic; i.e., the singular part of the Herglotz measure (relative to the Lebesque measure) is assumed to be absent. Capital P will indicate projection operators; $\vee$ will stand for 'the span of'; $-$ will be used for complex conjugation; $\sim$ will denote Hermitian transpose; $\oplus$ will be direct (or orthogonal) sum; $\otimes$ will indicate 'outer' or Kronecker product. Capital L will be reserved for index-sets, and $\mathrm{sym}\,^{m}L$ will denote a 'symmetric part' of the m-variate index-set $^{m}L = L \times \dots \times L$ (m-copies), obtained due to lexicographic ordering. The symbol $\mathbb{E}$ will stand for expectation.

2. NONLINEAR PREDICTION FILTER PROBLEM: A UNIFIED APPROACH

In this chapter we introduce the M-th degree nonlinear prediction filter problem for 2M-th order stochastic sequences observed on a finite time-interval. We state the problem algebraically and geometrically, in time- and in frequency-domains, showing that the nonlinear least-squares stochastic estimation problem can be equivalently considered as:

- the problem of deterministic approximation of the set of M-D impulse responses of the nonlinear filter, in the space of generalized (block, multi-indexed) matrices;
- the problem of deterministic approximation of the set of M-D transfer functions of that filter, in the space of generalized (block, multi-variate) z-polynomials.

2.1 Higher-order stochastic sequences

Let $\{y\}$ denote a scalar, zero-mean and purely stochastic, second-order Hilbert sequence observed on the time-interval $L = [0,\infty)$. This sequence will be represented by the random variables y_{-j} contained in the following column-matrix

$$Y = \text{col}\ [y_{-j}]_{j \in L} \tag{2.1}$$

The following two-indexed (second-order) covariance matrix is

$$H = \mathbb{E}\{Y \otimes \tilde{Y}\} = [\, h_{j,k} \,]_{j,k \in \mathbb{L}} \tag{2.2}$$

whose entries are given by

$$h_{j,k} = \mathbb{E}\{\, y_{-j}\bar{y}_{-k} \,\} \tag{2.3}$$

Given the covariance data (2.3), we can introduce the following 2-D Fourier transform pair

$$W(e^{i\theta}, e^{i\omega}) = \sum_{j=-\infty}^{\infty} \sum_{k=-\infty}^{\infty} h_{j,k}\, e^{-ij\theta}\, e^{ik\omega} \tag{2.4a}$$

$$h_{j,k} = \frac{1}{(2\pi)^2} \int d\theta \int d\omega \;\; e^{ij\theta}\, W(e^{i\theta}, e^{i\omega})\, e^{-ik\omega} \tag{2.4b}$$

where $W(e^{i\theta}, e^{i\omega})$ is the 2-D spectral function of $\{y\}$, satisfying

$$W(e^{i\theta}, e^{i\omega}) = \bar{W}(e^{i\omega}, e^{i\theta}) \tag{2.4c}$$

Let $n=0,\ldots,N$ and let $x=0,\ldots,N-n$. If the sequence is observed on the finite time-interval $L_n^x \triangleq \{x,\ldots,x+n\}$, where x is a reference point (offset) and n indicates the range (or 'order' of past observations), then Y will be replaced by the submatrix

$$Y_n^x = \mathrm{col}\,[\, y_{-j} \,]_{\, j \in L_n^x} \tag{2.5a}$$

and we will associate with $\{y\}$ the following Hermitian, positive-definite covariance matrix

$$H_n^x = \mathbb{E}\{Y_n^x \otimes \tilde{Y}_n^x\} = [h_{j,k}] \;\; (j,k) \in L_n^x \times L_n^x \tag{2.5b}$$

Assuming that the sequence is stationary (in a weak second-order sense), we will obtain for $j,k=\ldots,-1,0,1,\ldots$

$$h_{j,k} = h_{j+1,k+1} = h_{k-j} = r_p \quad , \quad p=k-j \tag{2.6a}$$

where $\{r_p\}$ is the covariance sequence. From (2.6a) it follows that

$$H_n^x = H_n^{x+1} \triangleq T_n = [h_{k-j}]_{(j,k) \in L_n^o \times L_n^o} \tag{2.6b}$$

regardless of the x-shift. We notice that (2.6) display the Toeplitz property of the two-indexed covariance matrix of the second-order stationary sequence $\{y\}$. In that case, the relations (2.4) will be replaced by the 1-D Fourier transforms, relating the spectral density (since the sequence is purely stochastic) to the covariance sequence.

Now let us consider a 2M-th order stochastic sequence $\{y\}$, satisfying for $j \in L$ and $m=1,\ldots,M$

$$\mathbb{E}\{ |y_{-j}|^{2m} \} < \infty \tag{2.7}$$

Then we can introduce the following M-block (column), m-indexed matrix of the random variables and their products (see also Appendix 1)

$${}^{\{M\}}Y = \mathrm{col}\ [{}^{m}Y]_{m=1,\ldots,M} \tag{2.8a}$$

whose block-entries are the following m-indexed matrices

$${}^{m}Y = [y_{-j_1} \cdots y_{-j_m}]_{j_1,\ldots,j_m \in L} \tag{2.8b}$$

We can associate with that sequence the generalized, $(M \times M)$-block (square), (m+u)-indexed covariance matrix

$$ {}^{\{M\times M\}}H = \mathbb{E}\{ {}^{\{M\}}Y \otimes {}^{\{M\}}\tilde{Y} \} = [{}^{m\oplus u}H]_{m,u=1,\dots,M} \tag{2.9a} $$

(where $\otimes$ indicates 'outer' product of generalized matrices, see also Appendix 1), whose (m+u)-indexed block-entries are expressed as

$$ {}^{m\oplus u}H = \mathbb{E}\{ {}^{m}Y \otimes {}^{u}\tilde{Y} \} = [h_{j_1\dots j_m k_1\dots k_u}]_{j_1,\dots,j_m,k_1,\dots,k_u \in L} \tag{2.9b} $$

with

$$ h_{j_1\dots j_m k_1\dots k_u} = \mathbb{E}\{ y_{-j_1}\dots y_{-j_m} \bar{y}_{-k_1}\dots \bar{y}_{-k_u} \} \tag{2.9c} $$

Each submatrix (2.9b) can therefore be interpreted as the (m+u)-th order covariance matrix of the 2M-th order sequence $\{y\}$. Let us observe that ${}^{\{M\times M\}}H$ will reduce to the two-indexed matrix H (2.2) of the second-order sequence if we assume $M=1$. We will show in the subsequent paragraphs of this chapter that the generalized covariance matrix (2.9) will be associated with the M-th degree nonlinear least-squares prediction problem, much like the two-indexed matrix (2.2) is associated with the linear problem. We notice that if the higher-order sequence is Gaussian, then for $m,u=1,\dots,M$

$$ \mathbb{E}\{y_{-j_1}\dots y_{-j_m}\bar{y}_{-k_1}\dots\bar{y}_{-k_u}\} = \begin{cases} \sum\prod \mathbb{E}\{y_{-j_r}\bar{y}_{-k_s}\} & , (m+u) \text{ even} \\ 0 & , (m+u) \text{ odd} \end{cases} \tag{2.10} $$

where symbol $\sum\prod$ stands for the summation over all distinct ways of partitioning the (m+u) random variables into products of averages of pairs, see e.g., Wiener (1958). This means that the 'information' about the Gaussian sequence is contained only the two-indexed submatrix ${}^{1\oplus 1}H$ of the ${}^{\{M\times M\}}H$, as the entries of all higher-order (i.e., even-order) covariance matrices ${}^{m\oplus u}H$ are expressible in terms of the entries of the left-upper submatrix ${}^{1\oplus 1}H$ (in the sense of (2.10)). Consequently, a nonlinear approach to the problem of optimum prediction of a 2M-th order stochastic sequence will yield better estimation accuracy if the sequence is non-Gaussian. Otherwise, the higher-order covariance matrices will not contain any 'new' information about the sequence, and the linear prediction filter (associated with ${}^{1\oplus 1}H$) will become the best possible filter (in the least-squares sense) for a Gaussian sequence.

Introducing a shorthand notation

$$e^{i\theta^m} = (e^{i\theta_1},\dots,e^{i\theta_m}) \quad ; \quad e^{i\omega^u} = (e^{i\omega_1},\dots,e^{i\omega_u}) \qquad (2.11)$$

we can consider for $m,u=1,\dots,M$ the M-D Fourier transform pair

$${}^{m\oplus u}W(e^{i\theta^m},e^{i\omega^u}) = \sum_{j_1=-\infty}^{\infty} \dots \sum_{j_m=-\infty}^{\infty} \sum_{k_1=-\infty}^{\infty} \dots \sum_{k_u=-\infty}^{\infty} h_{j_1\dots j_m k_1\dots k_u}$$

$$e^{-ij_1\theta_1}\dots e^{-ij_m\theta_m} e^{ik_1\omega_1}\dots e^{ik_u\omega_u} \qquad (2.12a)$$

$$h_{j_1\dots j_m k_1\dots k_u} = \frac{1}{(2\pi)^{m+u}} \int d\theta_1 \dots \int d\theta_m \int d\omega_1 \dots \int d\omega_u \; e^{ij_1\theta_1}\dots e^{ij_m\theta_m}$$

$${}^{m\oplus u}W(e^{i\theta^m},e^{i\omega^u})\; e^{-ik_1\omega_1}\dots e^{-k_u\omega_u} \qquad (2.12b)$$

The ${}^{m\oplus u}W(e^{i\theta^m},e^{i\omega^u})$ can be interpreted as the (m+u)-dimensional spectral function of the 2M-th order sequence, satisfying

$${}^{m\oplus u}W(e^{i\theta^m},e^{i\omega^u}) = {}^{m\oplus u}\bar{W}(e^{i\omega^u},e^{i\theta^m}) \qquad (2.13)$$

Then we can introduce the following spectral matrix

$$^{M\times M}W = [\,^{m\oplus u}W(e^{i\theta^m},e^{i\omega^u})]_{m,u=1,\dots,M} \tag{2.14}$$

containing precisely the same information about the 2M-th order sequence as the generalized covariance matrix $^{\{M\times M\}}H$. We notice that for a second-order sequence (corresponding to M=1) we would have

$$^{1\times 1}W = [\,^{1\oplus 1}W(e^{i\theta_1},e^{i\omega_1})] \tag{2.15}$$

$^{1\oplus 1}W$ being actually the two-dimensional spectral function (2.4a).

Let us introduce for $n=0,\dots,N$, $x=0,\dots,N-n$ and $m=1,\dots,M$ the following multi-variate index-sets

$$^{m}L_n^x \triangleq \{(j_1,\dots,j_m):\ j_r \in L_n^x\ ,\ r=1,\dots,m\} \tag{2.16a}$$

where $L_n^x = \{x,\dots,x+n\}$. By the 'symmetric part' $\mathrm{sym}\,^{m}L_n^x$ of the index-set $^{m}L_n^x$ we will understand

$$\mathrm{sym}\,^{m}L_n^x \triangleq \{(j_1,\dots,j_m):\ j_1 \in L_n^x\ ;\ j_r=j_{r-1},\dots,x+n\ ,\ r=2,\dots,m\} \tag{2.16b}$$

We notice that anti-lexicographic ordering can be equivalently introduced in (2.16b).

If the 2M-th order sequence $\{y\}$ is observed on a finite time-interval L_n^x, the $^{\{M\}}Y$ (2.8) will be replaced by

$$^{\{M\}}Y_n^x = \mathrm{col}\,[\,^{m}Y_n^x]_{m=1,\dots,M} \tag{2.17a}$$

whose block-entries are expressed as

$$^{m}Y_n^x = [\, y_{-j_1} \cdots y_{-j_m} \,]_{(j_1,\ldots,j_m) \in \, sym^m L_n^x} \qquad (2.17b)$$

where x is some reference point. Consequently, the generalized covariance matrix $^{\{M\times M\}}H$ (2.9) will become

$$^{\{M\times M\}}H_n^x = E\{^{\{M\}}Y_n^x \otimes {}^{\{M\}}\tilde{Y}_n^x\} = [\, ^{m\oplus u}H_n^x \,]_{m,u=1,\ldots,M} \qquad (2.18a)$$

with the block-entries given by

$$^{m\oplus u}H_n^x = E\{^{m}Y_n^x \otimes {}^{u}\tilde{Y}_n^x\} =$$

$$= [\, h_{j_1 \ldots j_m k_1 \ldots k_u} \,]_{(j_1,\ldots,j_m,k_1,\ldots,k_u) \in \, sym^m L_n^x \times sym^m L_n^x} \qquad (2.18b)$$

Assuming that $\{y\}$ is stationary (in a weak 2M-th order sense), we will have for $j_1,\ldots,j_m,k_1,\ldots,k_u = \ldots,-1,0,1,\ldots$

$$h_{j_1+1 \ldots j_m+1, k_1+1 \ldots k_u+1} = h_{j_1 \ldots j_m k_1 \ldots k_u} \qquad (2.19a)$$

so that

$$^{\{M\times M\}}H_n^x = {}^{\{M\times M\}}H_n^{x+1} \triangleq {}^{\{M\times M\}}T_n \qquad (2.19b)$$

regardless of the x-shift. Thus, (2.19) display the Toeplitz structure of the generalized covariance matrix of the stationary 2M-th order sequence. In that case, the block-Hankel, multi-indexed Hermitian, positive-definite matrix $^{\{M\times M\}}H$ will be replaced by the block-Hankel, multi-indexed Toeplitz, positive-definite matrix $^{\{M\times M\}}T$. We also notice that assuming M=1, (2.19) will reduce to (2.6).

2.2 Nonlinear least-squares prediction: Algebraic approach

Let us assume that the 2M-th order stochastic sequence $\{y\}$ is observed on the time-interval L_N^o and, hence, is represented by the random variables $y_o, y_{-1}, \ldots, y_{-N}$. In order to predict y_o in a nonlinear way, given its N-th order past $y_{-1}, \ldots, y_{-N}$, we define the M-th degree nonlinear, N-th order predictor ${}^M\hat{y}_{N;o}$ of y_o as the M-th degree Volterra functional polynomial

$$ {}^M\hat{y}_{N;o} \triangleq \sum_{m=1}^{M} \sum_{j_1=1}^{N} \sum_{j_2=j_1}^{N} \cdots \sum_{j_m=j_{m-1}}^{N} \alpha_{N;j_1\ldots j_m} \, y_{-j_1} \cdots y_{-j_m} \tag{2.20} $$

We notice that (2.20) will reduce to the linear N-th order estimate if we assume $M=1$ so that the linear approach will be the first step in the nonlinear treatment of the problem. The M-th degree nonlinear, N-th order prediction error will then be

$$ {}^M\varepsilon_{N;o} \triangleq y_o - {}^M\hat{y}_{N;o} \tag{2.21} $$

and the mean-square error, associated with (2.21), will be expressed as

$$ {}^M R_{\varepsilon;N} = \mathbb{E}\{ |{}^M\varepsilon_{N;o}|^2 \} \triangleq {}^M d_N^2 \geq 0 \tag{2.22} $$

The error (2.21) can be rewritten in a renormalized form as

$$ {}^M e_{N;o} = a_{N;o} y_o + \sum_{m=1}^{M} \sum_{j_1=1}^{N} \sum_{j_2=j_1}^{N} \cdots \sum_{j_m=j_{m-1}}^{N} a_{N;j_1\ldots j_m} \, y_{-j_1} \cdots y_{-j_m} \tag{2.23a} $$

where

$$ a_{N;o} = {}^M d_N^{-1} \quad ; \quad a_{N;j_1\ldots j_m} = - {}^M d_N^{-1} \, \alpha_{N;j_1\ldots j_m} \tag{2.23b} $$

In order to solve the M-th degree nonlinear, N-th order least-squares prediction problem, we wish to choose the coefficients $a_{N;j_1 \dots j_m}$ in a such way that the mean-square error (2.22) is minimized. This will be achieved if for $u=1,\dots,M$ and $(k_1,\dots,k_u) \in \text{sym}^u L^1_{N-1}$

$$\frac{\partial^M R_{\epsilon;N}}{\partial a_{N;k_1 \dots k_u}} = 0 \tag{2.24a}$$

Conditions (2.24a) will imply

$$a_{N;o} h_{o,k_1 \dots k_u} + \sum_{m=1}^{M} \sum_{j_1=1}^{N} \dots \sum_{j_m=j_{m-1}}^{N} a_{N;j_1 \dots j_m} h_{j_1 \dots j_m k_1 \dots k_u} = 0 \tag{2.24b}$$

Moreover, we get

$$a_{N;o} h_{oo} + \sum_{m=1}^{M} \sum_{j_1=1}^{N} \dots \sum_{j_m=j_{m-1}}^{N} a_{N;j_1 \dots j_m} h_{j_1 \dots j_m, o} = {}^M d_N$$

Generalizing the notion of 'matrix', we can introduce the following M-block (row), m-indexed coefficient-matrix

$$\{M\} A_N = [\, {}^m A_N \,]_{m=1,\dots,M} \tag{2.25a}$$

whose each m-indexed block-entry ${}^m A_N$ can be considered as a map from the m-variate index-set $D\,{}^m A_N$ into the real (or complex) numbers

$${}^m A_N : D\,{}^m A_N \to \mathbb{R} \tag{2.25b}$$

Hence, we can talk about the 'domain' of the m-indexed matrix as being that m-variate index-set, and we will have

$$D\,{}^m A_N = \begin{cases} L^o_N & \text{if } m=1 \\ \text{sym}^m L^1_{N-1} & \text{if } m=2,\dots,M \end{cases} \tag{2.25c}$$

Consequently, we write for $m=2,\dots,M$

$$ {}^{m}A_{N} = [\, a_{N;j_1\dots j_m} \,]_{(j_1,\dots,j_m)} \in \mathrm{sym}\,{}^{m}L^{1}_{N-1} \tag{2.25d} $$

and for $m=1$

$$ {}^{1}A_{N} = [\, a_{N;j_1} \,]_{j_1 \in L^{o}_{N}} \tag{2.25e} $$

Thus, the M-block, m-indexed coefficient-matrix ${}^{\{M\}}A_N$ (2.25a) can be considered as a map from the vector of simple domains

$$ D\,{}^{\{M\}}A_{N} = [D\,{}^{m}A_{N}]_{m=1,\dots,M} \tag{2.26a} $$

into the real (or complex) numbers

$$ {}^{\{M\}}A_{N} : D\,{}^{\{M\}}A_{N} \rightarrow \mathbb{R} \tag{2.26b} $$

see also Appendix 1.

Using (2.25),(2.26) and introducing

$$ {}^{\{M\}}Y_{N} = \mathrm{col}\,[\, {}^{1}Y^{o}_{N} \quad {}^{2}Y^{1}_{N-1} \quad \dots \quad {}^{M}Y^{1}_{N-1} \,] \tag{2.27a} $$

where ${}^{1}Y^{o}_{N}$ is expressed by (2.17b) with $m=1$, $x=0$ and $n=N$ while ${}^{m}Y^{1}_{N-1}$ $(m=2,\dots,M)$ are given by (2.17b) with $x=1$ and $n=N-1$, we can rewrite the error (2.23) in a generalized matrix form as follows

$$ {}^{M}e_{N;o} = {}^{\{M\}}A_{N} \bullet {}^{\{M\}}Y_{N} \tag{2.27b} $$

where $\bullet$ indicates product of generalized matrices (see Appendix 1).

Introducing the following matrix

$$[\ {}^{M}d_N \quad {}^{\{M\}}O^1_{N-1}] \tag{2.28a}$$

where ${}^{M}d_N$ is expressed by (2.22), and ${}^{\{M\}}O^1_{N-1}$ denotes the M-block (row), u-indexed zero-matrix

$${}^{\{M\}}O^1_{N-1} = [{}^{u}O^1_{N-1}]_{\ u=1,\dots,M} \tag{2.28b}$$

whose block-entries ${}^{u}O^1_{N-1}$ are u-indexed zero-matrices with domains $D\,{}^{u}O^1_{N-1} = \mathrm{sym}\,{}^{u}L^1_{N-1}$ (see also Appendix 1), we can rewrite the 'normal equations' (2.24) in a generalized matrix form as follows (see Zarzycki and Dewilde (1983a))

$${}^{\{M\}}A_N \cdot {}^{\{M\times M\}}H_N = [\ {}^{M}d_N \quad {}^{\{M\}}O^1_{N-1}\ ] \tag{2.29}$$

where ${}^{\{M\times M\}}H_N$ is expressed by (2.18) with ${}^{\{M\}}Y^x_n$ replaced by ${}^{\{M\}}Y_N$ (2.27a). Rewriting ${}^{\{M\times M\}}H_N$ in a generalized block-column form

$${}^{\{M\times M\}}H_N = [\ {}^{\{M\}\times u}H_N\ ]_{\ u=1,\dots,M} \tag{2.30a}$$

where for $u=2,\dots,M$

$${}^{\{M\}\times u}H_N = [\ {}^{\{M\}}H_{N;k_1\dots k_u}\ ]_{\ (k_1,\dots,k_u)\in\,\mathrm{sym}\,{}^{u}L^1_{N-1}} \tag{2.30b}$$

and

$${}^{\{M\}\times 1}H_N = [\ {}^{\{M\}}H_{N;k_1}\ ]_{\ k_1\in L^o_N} \tag{2.30c}$$

with

$${}^{\{M\}}H_{N;k_1\dots k_u} = [h_{j_1\dots j_m k_1\dots k_u}]_{\ (j_1,\dots,j_m)\in D\,{}^{\{M\}}Y_N} \tag{2.30d}$$

we can express the 'normal equations' (2.29), for $u=1,\dots,M$ and for $(k_1,\dots,k_u) \in sym^{u}L^{1}_{N-1}$, as follows

$$^{\{M\}}A_N \cdot {}^{\{M\}}H_{N;k_1\dots k_u} = 0 \qquad (2.31a)$$

$$^{\{M\}}A_N \cdot {}^{\{M\}}H_{N;o} = {}^{M}d_N \qquad (2.31b)$$

These are the 'normal equations' associated with the M-th degree nonlinear, N-th order prediction problem. We can observe that they will reduce to the 'normal equations' corresponding to the linear problem, if $M=1$.

Equations (2.31) can be recursively solved via the nonlinear Levinson algorithm, presented in Zarzycki and Dewilde (1983a); Zarzycki (1983). This algorithm computes the coefficients of the optimum nonlinear approximate prediction ladder-filter, and implies the generalized Cholesky factorization of the block, multi-indexed covariance matrix $^{\{M\times M\}}H_N$, generalizing the linear case, considered in Deprettere and Lie (1980).

2.3 Nonlinear least-squares prediction: Geometric approach

In the subsequent sections of this paragraph we will introduce the nonlinear prediction problem in the following spaces: of the regular Volterra functional polynomials; of the generalized coefficient-matrices; and of generalized z-polynomials.

2.3.1 Space of the regular Volterra functional polynomials

Given the M-block, m-indexed matrix of the random variables $^{\{M\}}Y^{1}_{N-1}$ (expressed by (2.17) with $x=1$ and $n=N-1$), and assuming that the entries are linearly independent, we can consider the space

$$ {}^{\{M\}}\mathbf{Y}^{1}_{N-1} \stackrel{\Delta}{=} \vee\{{}^{\{M\}}Y^{1}_{N-1}\} \tag{2.32} $$

Each element from that space will be expressed as

$$ {}^{M}\varphi_{N} = {}^{\{M\}}F_{N}\cdot{}^{\{M\}}Y^{1}_{N-1} \tag{2.33a} $$

where the M-block (row), m-indexed coefficient-matrix is given by

$$ {}^{\{M\}}F_{N} = [{}^{m}F_{N}]_{m=1,\dots,M} \tag{2.33b} $$

with

$$ {}^{m}F_{N} = [f_{N;j_1\dots j_m}]_{(j_1,\dots,j_m)\in \mathrm{sym}\,{}^{m}L^{1}_{N-1}} \tag{2.33c} $$

From Appendix 1 it follows that (2.33) can be rewritten as

$$ {}^{M}\varphi_{N} = \sum_{m=1}^{M} {}^{m}F_{N}\cdot{}^{m}Y^{1}_{N-1} \tag{2.33d} $$

Thus, we can observe that ${}^{M}\varphi_{N}$ is the regular M-th degree Volterra functional polynomial of the form (1.2). We also notice that assuming $M=1$ we obtain the space ${}^{\{1\}}\mathbf{Y}^{1}_{N-1}$, associated with the linear N-th order prediction problem. Given two Volterra functional polynomials ${}^{M}\varphi_{N}$ and ${}^{M}\psi_{N}$ (where the latter element will be expressed by (2.33) with F and f replaced by G and g, respectively), we can introduce inner-product on ${}^{\{M\}}\mathbf{Y}^{1}_{N-1}$ as

$$ ({}^{M}\varphi_{N}, {}^{M}\psi_{N})_{\mathbf{Y}} \stackrel{\Delta}{=} \mathbf{E}\{{}^{M}\varphi_{N}\,{}^{M}\bar{\psi}_{N}\} \tag{2.34a} $$

inducing the norm

$$\|{}^{M}\varphi_N\|^2_{\mathbf{Y}} = \mathbb{E}\{|{}^{M}\varphi_N|^2\} \tag{2.34b}$$

2.3.2 Space of generalized coefficient-matrices

Let us introduce for $m=1,\dots,M$ and $(j_1,\dots,j_m) \in sym^m L^1_{N-1}$ the following M-block, m-indexed matrices

$$^{\{M\}}1_{j_1\dots j_m} \triangleq [{}^{1}O^1_{N-1} \; \dots \; {}^{m-1}O^1_{N-1} \; {}^{m}1_{j_1\dots j_m} \; {}^{m+1}O^1_{N-1} \; \dots \; {}^{M}O^1_{N-1}] \tag{2.35a}$$

where each ${}^{u}O^1_{N-1}$ is the u-indexed zero-matrix of the form (2.28b); the m-indexed matrix ${}^{m}1_{j_1\dots j_m}$ is given by

$$^{m}1_{j_1\dots j_m} = [\delta_{j_1\dots j_m;k_1\dots k_m}]_{(k_1,\dots,k_m) \in sym^m L^1_{N-1}} \tag{2.35b}$$

with $\delta_{j_1\dots j_m;k_1\dots k_m}$ denoting the m-dimensional Kronecker delta

$$\delta_{j_1\dots j_m;k_1\dots k_m} = \begin{cases} 1 & \text{if } k_1=j_1, \dots, k_m=j_m \\ 0 & \text{otherwise} \end{cases} \tag{2.35c}$$

This means that ${}^{m}1_{j_1\dots j_m}$ and, hence, ${}^{\{M\}}1_{j_1\dots j_m}$ contains only one unit-entry with coordinates $(j_1,\dots,j_m)$, placed in the m-indexed block, while all remaining entries are equal to zero. Now let for $m=1,\dots,M$

$$^{m}I^1_{N-1} = [{}^{\{M\}}1_{j_1\dots j_m}]_{(j_1,\dots,j_m) \in sym^m L^1_{N-1}} \tag{2.36a}$$

and, finally, let

$$^{\{M\}}I^1_{N-1} = [{}^{m}I^1_{N-1}]_{m=1,\dots,M} \tag{2.36b}$$

We can observe that the entries ${}^{\{M\}}1_{j_1\dots j_m}$ of the ${}^{\{M\}}I^1_{N-1}$ are linearly independent, and they can be considered as the 'natural' basis of the space

$$ {}^{\{M\}}\mathbb{I}^1_{N-1} \triangleq \vee \{ {}^{\{M\}}I^1_{N-1} \} \tag{2.37} $$

whose elements will be expressed as

$$ {}^{\{M\}}F_N = \sum_{m=1}^{M} \sum_{j_1=1}^{N} \sum_{j_2=j_1}^{N} \dots \sum_{j_m=j_{m-1}}^{N} f_{N;j_1\dots j_m} \; {}^{\{M\}}1_{j_1\dots j_m} = $$

$$ = [\, {}^{m}F_N \,]_{\; m=1,\dots,M} \tag{2.38} $$

where ${}^{m}F_N$ is expressed by (2.33c). This means that ${}^{\{M\}}\mathbb{I}^1_{N-1}$ is the space of M-block, m-indexed coefficient-matrices.

Given the generalized covariance matrix ${}^{\{M\times M\}}H^1_{N-1}$ (expressed by (2.18) with $x=1$ and $n=N-1$) of the 2M-th order stochastic sequence $\{y\}$, we define the weighted inner-product on ${}^{\{M\}}\mathbb{I}^1_{N-1}$ as

$$ ({}^{\{M\}}F_N, {}^{\{M\}}G_N)_{\mathbb{I}} \triangleq {}^{\{M\}}F_N \cdot {}^{\{M\times M\}}H^1_{N-1} \cdot {}^{\{M\}}\tilde{G}_N \tag{2.39a} $$

where ${}^{\{M\}}G_N$ is expressed by (2.38) with F and f replaced by G and g, respectively. Recalling that the generalized covariance matrix is multi-indexed Hermitian and positive-definite, we can check that

$$ ({}^{\{M\}}F_N, {}^{\{M\}}G_N)_{\mathbb{I}} = ({}^{\{M\}}G_N, {}^{\{M\}}F_N)^{\sim}_{\mathbb{I}} \tag{2.39b} $$

$$ (a\cdot {}^{\{M\}}A_N + b\cdot {}^{\{M\}}B_N, {}^{\{M\}}F_N)_{\mathbb{I}} = a({}^{\{M\}}A_N, {}^{\{M\}}F_N)_{\mathbb{I}} + b({}^{\{M\}}B_N, {}^{\{M\}}F_N)_{\mathbb{I}} \tag{2.39c} $$

$$({}^{\{M\}}F_N, {}^{\{M\}}F_N)_{\mathbb{I}} = {}^{\{M\}}F_N \cdot {}^{\{M\times M\}}H^1_{N-1} \cdot {}^{\{M\}}\tilde{F}_N \geq 0 \qquad (2.39d)$$

with equality in (2.39d) iff ${}^{\{M\}}F_N = {}^{\{M\}}O^1_{N-1}$. The norm is then

$$\| {}^{\{M\}}F_N \|^2_{\mathbb{I}} = ({}^{\{M\}}F_N, {}^{\{M\}}F_N)_{\mathbb{I}} \qquad (2.39e)$$

We can observe that

$$\| {}^{\{M\}}1_{j_1\ldots j_m} \|^2_{\mathbb{I}} = {}^{\{M\}}1_{j_1\ldots j_m} \cdot {}^{\{M\times M\}}H^1_{N-1} \cdot {}^{\{M\}}\tilde{1}_{j_1\ldots j_m} =$$

$$= h_{j_1 j_1 \ldots j_m j_m} > 0 \qquad (2.40a)$$

We also notice that the 'natural' basis of the space ${}^{\{M\}}\mathbb{I}^1_{N-1}$ is not, in general, an orthogonal basis of that space, as we have for $m,u=1,\ldots,M$ and $(j_1,\ldots,j_m,k_1,\ldots,k_u) \in \mathrm{sym}^m L^1_{N-1} \times \mathrm{sym}^m L^1_{N-1}$

$$({}^{\{M\}}1_{j_1\ldots j_m}, {}^{\{M\}}1_{k_1\ldots k_u})_{\mathbb{I}} = h_{j_1\ldots j_m k_1 \ldots k_u} \qquad (2.40b)$$

If the underlying 2M-th order sequence was Gaussian, we would have (in accordance with (2.10))

$$^{m\oplus u}H^1_{N-1} = {}^{m\oplus u}O^1_{N-1} \quad , \quad (m+u) \text{ odd} \qquad (2.41)$$

Consequently, for $(m+u)$ odd,

$$({}^{\{M\}}1_{j_1\ldots j_m}, {}^{\{M\}}1_{k_1\ldots k_u})_{\mathbb{I}} = 0 \qquad (2.42a)$$

and we would have the 'block' orthogonality of the 'natural' basis in the Gaussian case. Moreover, if the sequence was white, we would write

$$({}^{\{M\}}1_{j_1\ldots j_m}, {}^{\{M\}}1_{k_1\ldots k_m})_{\mathbb{I}} = \Sigma\,\Pi\,\delta_{j_r,k_s} \qquad (2.42b)$$

in accordance with (2.10). In general, if

$$\left({}^{\{M\}}F_N, {}^{\{M\}}G_N\right)_{\mathrm{II}} = 0 \tag{2.43}$$

we shall say that two M-block (row), m-indexed coefficient-matrices are orthogonal with respect to the generalized block, multi-indexed covariance matrix of the underlying 2M-th order stochastic sequence $\{y\}$. If the sequence was stationary (in a weak 2M-th order sense), the block, multi-indexed Toeplitz covariance matrix ${}^{\{M\times M\}}T_{N-1}$ would replace ${}^{\{M\times M\}}H$ in (2.39), in accordance with (2.19).

2.3.3 Space of generalized z-polynomials

Let us introduce for $m=1,\ldots,M$ and $(j_1,\ldots,j_m) \in \mathrm{sym}\,{}^mL^1_{N-1}$ the following row-matrices

$$
\begin{aligned}
{}^{M}z_{j_1\ldots j_m} &= [\,\delta_{m,p}\cdot z_1^{j_1}\ldots z_m^{j_m}\,]_{p=1,\ldots,M} = \\
&= [0 \;\ldots\; 0 \;\; z_1^{j_1}\ldots z_m^{j_m} \;\; 0 \;\ldots\; 0]
\end{aligned} \tag{2.43a}
$$

and let

$$ {}^{m\times M}Z^1_{N-1} = [\,{}^{M}z_{j_1\ldots j_m}\,]_{(j_1,\ldots,j_m)\in \mathrm{sym}\,{}^mL^1_{N-1}} \tag{2.43b}$$

so that we can introduce

$$ {}^{\{M\}}Z^1_{N-1} = [\,{}^{m\times M}Z^1_{N-1}\,]_{m=1,\ldots,M} \tag{2.43c}$$

Then we can consider the following space

$$ {}^{\{M\}}\mathbf{Z}^1_{N-1} \triangleq \vee\{\,{}^{\{M\}}Z^1_{N-1}\,\} \tag{2.44}$$

whose elements will be expressed as

$$ {}^{M}\Phi_N = \sum_{m=1}^{M} \sum_{j_1=1}^{N} \sum_{j_2=j_1}^{N} \cdots \sum_{j_m=j_{m-1}}^{N} f_{N;j_1\ldots j_m} \; {}^{M}z_{j_1\ldots j_m} = $$

$$ = \sum_{m=1}^{M} [0 \;\ldots\; 0 \;\; \Phi_N(z_1,\ldots,z_m) \;\; 0 \;\ldots\; 0] = $$

$$ = [\Phi_N(z_1,\ldots,z_m)]_{m=1,\ldots,M} \tag{2.45a} $$

where $\Phi_N(z_1,\ldots,z_m)$ denotes the following m-variate z-polynomial

$$ \Phi_N(z_1,\ldots,z_m) = \sum_{j_1=1}^{N} \sum_{j_2=1}^{N} \cdots \sum_{j_m=j_{m-1}}^{N} f_{N;j_1\ldots j_m} \; z_1^{j_1} \ldots z_m^{j_m} \tag{2.45b} $$

Thus, ${}^{\{M\}}\mathbf{z}_{N-1}^{1}$ is the space of the M-block, m-variate z-polynomials. Given two elements ${}^{M}\Phi_N$ and ${}^{M}\Psi_N$ from that space (where the latter is given by (2.45) with F and f replaced by G and g , respectively), and using the spectral matrix (2.14) of the underlying 2M-th order sequence {y}, we introduce inner-product on ${}^{\{M\}}\mathbf{z}_{N-1}^{1}$ as follows

$$ ({}^{M}\Phi_N, {}^{M}\Psi_N)_{\mathbf{z}} = \int d\theta^{\{M\}} \int d\omega^{\{M\}} \, {}^{M}\Phi_N \cdot {}^{M\times M}W \cdot {}^{M}\tilde{\Psi}_N = $$

$$ \triangleq \sum_{m=1}^{M} \sum_{u=1}^{M} \frac{1}{(2\pi)^{m+u}} \int d\theta^m \int d\omega^u \; \Phi_N(z_1,\ldots,z_m) $$

$$ {}^{m\oplus u}W(e^{i\theta^m}, e^{i\omega^u}) \times \bar{\Psi}_N(w_1,\ldots,w_u) \Bigg|_{\substack{z_r = e^{i\theta_r}, \; r=1,\ldots,m \\ w_s = e^{i\omega_s}, \; s=1,\ldots,u}} \tag{2.46a} $$

where we used the following shorthand notation

$$ \int d\theta^m = \int d\theta_1 \ldots \int d\theta_m \quad ; \quad \int d\omega^u = \int d\omega_1 \ldots \int d\omega_u \tag{2.46b} $$

Using (2.13), we can show that

$$({}^{M}\Phi_N, {}^{M}\Psi_N)_{\mathbf{Z}} = \sum_{m=1}^{M} \sum_{u=1}^{M} \frac{1}{(2\pi)^{m+u}} \int d\theta^m \int d\omega^u \; \overline{\Psi_N(w_1,\ldots,w_u)}$$

$$\overline{{}^{m\oplus u}W(e^{i\omega^u}, e^{i\theta^m}) \times \bar{\Phi}_N(z_1,\ldots,z_m)} \Bigg|_{\substack{z_r = e^{i\theta_r},\; r=1,\ldots,m \\ w_s = e^{i\omega_s},\; s=1,\ldots,u}} =$$

$$= \overline{({}^{M}\Psi_N, {}^{M}\Phi_N)_{\mathbf{Z}}} \tag{2.46c}$$

We also have

$$(a \cdot {}^{M}\Gamma_N + b \cdot {}^{M}\Delta_N, {}^{M}\Phi_N)_{\mathbf{Z}} = a({}^{M}\Gamma_N, {}^{M}\Phi_N)_{\mathbf{Z}} + b({}^{M}\Delta_N, {}^{M}\Phi_N)_{\mathbf{Z}} \tag{2.46d}$$

and, finally,

$$({}^{M}\Phi_N, {}^{M}\Phi_N)_{\mathbf{Z}} = \| {}^{M}\Phi_N \|_{\mathbf{Z}}^2 \geq 0 \tag{2.46e}$$

If $({}^{M}\Phi_N, {}^{M}\Psi_N)_{\mathbf{Z}} = 0$, we shall say that two generalized (M-block, m-variate) z-polynomials are orthogonal with respect to the spectral matrix ${}^{M\times M}W$ (2.14) of the 2M-th order sequence, on the unit (m+u)-torus.

REMARK: Let $\mathbf{T}$: $\{e^{i\theta}\}$ denote the unit-circle. Then for the unit m-torus ${}^{m}\mathbf{T}$ we will have

$${}^{m}\mathbf{T} = \underbrace{\mathbf{T} \otimes \ldots \otimes \mathbf{T}}_{m} : \underbrace{\{e^{i\theta}\} \otimes \ldots \otimes \{e^{i\theta}\}}_{m} = \{e^{i\theta_1} \ldots e^{i\theta_m}\}$$

while for the unit (m+u)-torus ${}^{m\oplus u}\mathbf{T}$ we get

$${}^{m\oplus u}\mathbf{T} = {}^{m}\mathbf{T} \otimes {}^{u}\mathbf{T} : \{e^{i\theta_1} \ldots e^{i\theta_m}\} \otimes \{e^{i\theta_1} \ldots e^{i\theta_u}\} =$$

$$= \{e^{i\theta_1} \ldots e^{i\theta_m} \, e^{i\omega_1} \ldots e^{i\omega_u}\}$$

2.3.4 Isometries

We can observe that ${}^{\{M\}}\mathbf{Y}^{1}_{N-1}$, ${}^{\{M\}}\mathbf{I}^{1}_{N-1}$ and ${}^{\{M\}}\mathbf{Z}^{1}_{N-1}$ are isomorphic spaces; i.e., we have

$$\begin{array}{ccc} & y_{-j_1}\cdots y_{-j_m} & \\ \swarrow\!\!\nearrow & & \nwarrow\!\!\searrow \\ {}^{\{M\}}1_{j_1\cdots j_m} & \leftrightarrow & {}^{M}z_{j_1\cdots j_m} \end{array} \tag{2.47a}$$

so that

$$\begin{array}{ccc} & {}^{M}\varphi_N & \\ \swarrow\!\!\nearrow & & \nwarrow\!\!\searrow \\ {}^{\{M\}}F_N & \leftrightarrow & {}^{M}\Phi_N \end{array} \tag{2.47b}$$

We wish to show that the following isometries hold

$$\begin{array}{ccc} & (y_{-j_1}\cdots y_{-j_m},\, y_{-k_1}\cdots y_{-k_u})_{\mathbf{Y}} & \\ /\!/ & & \backslash\!\backslash \\ ({}^{\{M\}}1_{j_1\cdots j_m},\, {}^{\{M\}}1_{k_1\cdots k_u})_{\mathbf{I}} & = & ({}^{M}z_{j_1\cdots j_m},\, {}^{M}z_{k_1\cdots k_u})_{\mathbf{Z}} \end{array} \tag{2.47c}$$

so that

$$\begin{array}{ccc} & ({}^{M}\varphi_N,\, {}^{M}\psi_N)_{\mathbf{Y}} & \\ /\!/ & & \backslash\!\backslash \\ ({}^{\{M\}}F_N,\, {}^{\{M\}}G_N)_{\mathbf{I}} & = & ({}^{M}\Phi_N,\, {}^{M}\Psi_N)_{\mathbf{Z}} \end{array} \tag{2.47d}$$

In order to do that, let us observe (following (2.9c),(2.34a) and (2.40b)) that

$$(y_{-j_1}\cdots y_{-j_m}, y_{-k_1}\cdots y_{-k_u})_{\mathbf{Y}} = \mathbb{E}\{ y_{-j_1}\cdots y_{-j_m}\bar{y}_{-k_1}\cdots\bar{y}_{-k_u}\} =$$

$$= h_{j_1\cdots j_m k_1\cdots k_u} =$$

$$= ({}^{\{M\}}1_{j_1\cdots j_m}, {}^{\{M\}}1_{k_1\cdots k_u})_{\mathbb{I}} \qquad (2.48a)$$

From (2.11b) and (2.46a) we obtain

$$({}^{M}z_{j_1\cdots j_m}, {}^{M}z_{k_1\cdots k_u})_{\mathbf{Z}} = \frac{1}{(2\pi)^{m+u}} \int d\theta^m \int d\omega^u \; z_1^{j_1}\cdots z_m^{j_m} \times$$

$$\times \; {}^{m\oplus u}W(e^{i\theta^m}, e^{i\omega^u}) \; w_1^{-k_1}\cdots w_u^{-k_u}\Bigg|_{\substack{z_r = e^{i\theta_r}, \; r=1,\dots,m \\ w_s = e^{i\omega_s}, \; s=1,\dots,u}} =$$

$$= h_{j_1\cdots j_m k_1\cdots k_u} =$$

$$= (y_{-j_1}\cdots y_{-j_m}, y_{-k_1}\cdots y_{-k_u})_{\mathbf{Y}} = ({}^{\{M\}}1_{j_1\cdots j_m}, {}^{\{M\}}1_{k_1\cdots k_u})_{\mathbb{I}} \qquad (2.48b)$$

so that (2.47c) is valid. From (2.34a),(2.39a) and (2.48a) it follows that

$$({}^{M}\varphi_N, {}^{M}\psi_N)_{\mathbf{Y}} = \mathbb{E}\{{}^{M}\varphi_N \, {}^{M}\bar{\psi}_N\} = \sum_{m=1}^{M}\sum_{u=1}^{M} {}^{m}F_N \cdot {}^{m\oplus u}H^1_{N-1} \cdot {}^{u}\tilde{G}_N =$$

$$= {}^{\{M\}}F_N \cdot {}^{\{M\times M\}}H^1_{N-1} \cdot {}^{\{M\}}\tilde{G}_N =$$

$$= ({}^{\{M\}}F_N, {}^{\{M\}}G_N)_{\mathbb{I}} \qquad (2.49a)$$

Finally, using (2.12b),(2.34a),(2.45b),(2.46a) and (2.48b), we obtain

$$({}^{M}\Phi_N, {}^{M}\Psi_N)_{\mathbf{Z}} = \sum_{m=1}^{M} \sum_{u=1}^{M} \frac{1}{(2\pi)^{m+u}} \int d\theta^m \int d\omega^u \; \Phi_N(z_1,\dots,z_m) \times$$

$$\times \; {}^{m\oplus u}W(e^{i\theta^m}, e^{i\omega^u}) \; \bar{\Psi}_N(w_1,\dots,w_u) \Bigg|_{\substack{z_r = e^{i\theta_r}, \; r=1,\dots,m \\ w_s = e^{i\omega_s}, \; s=1,\dots,u}} =$$

$$= \sum_{m=1}^{M} \sum_{u=1}^{M} \sum_{j_1=1}^{N} \dots \sum_{j_m=j_{m-1}}^{N} \sum_{k_1=1}^{N} \dots \sum_{k_u=k_{u-1}}^{N} f_{N;j_1\dots j_m} \times$$

$$\times \frac{1}{(2\pi)^{m+u}} \int d\theta^m \int d\omega^u \; e^{ij_1\theta_1} \dots e^{ij_m\theta_m} \; {}^{m\oplus u}W(e^{i\theta^m}, e^{i\omega^u})$$

$$e^{-ik_1\omega_1} \dots e^{-ik_u\omega_u} \; \bar{g}_{N;k_1\dots k_u} =$$

$$= \sum_{m=1}^{M} \sum_{u=1}^{M} \sum_{j_1=1}^{N} \dots \sum_{j_m=j_{m-1}}^{N} \sum_{k_1=1}^{N} \dots \sum_{k_u=k_{u-1}}^{N} f_{N;j_1\dots j_m}$$

$$h_{j_1\dots j_m k_1\dots k_u} \; \bar{g}_{N;k_1\dots k_u} =$$

$$= \sum_{m=1}^{M} \sum_{u=1}^{M} {}^{m}F_N \cdot {}^{m\oplus u}H^1_{N-1} \cdot {}^{u}\tilde{G}_N =$$

$$= {}^{\{M\}}F_N \cdot {}^{\{M\times M\}}H^1_{N-1} \cdot {}^{\{M\}}\tilde{G}_N =$$

$$= ({}^{M}\varphi_N, {}^{M}\psi_N)_{\mathbf{Y}} = ({}^{\{M\}}F_N, {}^{\{M\}}G_N)_{\mathbf{I}} \qquad (2.49b)$$

This proves (2.47d).

From (2.47) it follows that the M-th degree nonlinear, N-th order stochastic estimation problem considered in the space ${}^{\{M\}}\mathbf{Y}^1_{N-1}$, can be equivalently interpreted as the deterministic problem in the space ${}^{\{M\}}\mathbf{I}^1_{N-1}$

of generalized matrices, or as the deterministic problem in the space ${}^{\{M\}}\mathbf{Z}^{1}_{N-1}$ of generalized z-polynomials. The isomorphism (2.47) is a generalization of the isomorphism considered in the linear least-squares estimation problem (which can be immediately obtained from (2.47), assuming $M=1$), and becomes the Kolmogorov isomorphism in stationary case.

Now we are in a position to show that the nonlinear stochastic estimation problem can be equivalently considered as the problem of (deterministic) optimum approximation of the set of M-D impulse responses and/or transfer functions of the nonlinear prediction filter.

2.3.5 Stochastic nonlinear estimation

Given the space ${}^{\{M\}}\mathbf{Y}^{1}_{N-1}$ (2.32), let us introduce

$$ {}^{\{M\}}\mathbf{Y}_{N} = \vee\{\, y_o \,,\ {}^{\{M\}}\mathbf{Y}^{1}_{N-1} \} \tag{2.50}$$

The M-th degree nonlinear, N-th order least-squares predictor ${}^{M}\hat{y}_{N;o}$ of y_o will be expressed as

$$ {}^{M}\hat{y}_{N;o} \triangleq {}^{\{M\}}P_{\mathbf{Y};N-1}^{\;1}\, y_o \ \in\ {}^{\{M\}}\mathbf{Y}^{1}_{N-1} \tag{2.51a}$$

where ${}^{\{M\}}P_{\mathbf{Y};N-1}^{\;1}$ is the orthogonal projection operator, taking projection on the subspace ${}^{\{M\}}\mathbf{Y}^{1}_{N-1}$. We notice that ${}^{M}\hat{y}_{N;o}$ is precisely the nonlinear estimate (2.20), introduced algebraically. The M-th degree nonlinear, N-th order prediction error, corresponding to ${}^{M}\hat{y}_{N;o}$, is given by

$$ {}^{M}\varepsilon_{N;o} \triangleq {}^{\{M\}}P_{\mathbf{Y};N-1}^{\perp\,1}\, y_o = y_o - {}^{M}\hat{y}_{N;o} \ \perp\ {}^{\{M\}}\mathbf{Y}^{1}_{N-1} \tag{2.51b}$$

and can be expressed in a renormalized form as

$$^{M}e_{N;o} = {}^{M}\varepsilon_{N;o} \, \| {}^{M}\varepsilon_{N;o} \|_{\mathbf{Y}}^{-1} = {}^{\{M\}}A_N \cdot {}^{\{M\}}\mathbf{Y}_N \tag{2.51c}$$

where ${}^{\{M\}}A_N$ and ${}^{\{M\}}\mathbf{Y}_N$ are given by (2.25) and (2.27a), respectively. We notice that this is precisely the nonlinear normalized prediction error, introduced algebraically in (2.27b).

The orthogonality conditions (2.51b) imply for $u=1,\ldots,M$ and $(k_1,\ldots,k_u) \in \mathrm{sym}\,{}^{u}L_{N-1}^{1}$

$$({}^{M}e_{N;o}\,,\, y_{-k_1} \cdots y_{-k_u})_{\mathbf{Y}} = 0 \tag{2.52a}$$

Moreover, the norm of the error will be expressed as

$$({}^{M}e_{N;o}\,,\, y_o)_{\mathbf{Y}} = \| {}^{M}\varepsilon_{N;o} \|_{\mathbf{Y}} \tag{2.52b}$$

These are the 'normal equations', associated with the M-th degree nonlinear least-squares prediction problem introduced in the space of the regular Volterra functional polynomials. We can observe that conditions (2.52) are precisely the 'normal equations' (2.31), derived algebraically. Moreover, we remark that the 'normal equations' for the linear N-th order predictor are given by (2.52) with $M=1$.

2.3.6 Optimum generalized matrix approximation

Considering the space ${}^{\{M\}}\mathbf{I}_{N-1}^{1}$ (2.37), we define

$${}^{\{M\}}\mathbf{I}_N = \vee \{ {}^{\{M\}}1_o \,,\, {}^{\{M\}}I_{N-1}^{1} \} \tag{2.53}$$

The N-th order estimate ${}^{\{M\}}\hat{1}_{N;o}$ of the element ${}^{\{M\}}1_o$ will be introduced as follows

$$ {}^{\{M\}}\hat{1}_{N;o} \triangleq {}^{\{M\}}P^{1}_{\mathbb{I};N-1}\;{}^{\{M\}}1_{o} \in {}^{\{M\}}\mathbb{I}^{1}_{N-1} \tag{2.54a} $$

where ${}^{\{M\}}P^{1}_{\mathbb{I};N-1}$ indicates the orthogonal projection operator on the subspace ${}^{\{M\}}\mathbb{I}^{1}_{N-1}$. The N-th order approximation error is then given by

$$ {}^{\{M\}}A_{N} \triangleq {}^{\{M\}}P^{\perp\,1}_{\mathbb{I};N-1}\;{}^{\{M\}}1_{o} = {}^{\{M\}}1_{o} - [\,0\;\; {}^{\{M\}}\hat{1}_{N;o}\,] \perp {}^{\{M\}}\mathbb{I}^{1}_{N-1} \tag{2.54b} $$

and can be rewritten in a renormalized form as

$$ {}^{\{M\}}A_{N} = {}^{\{M\}}A_{N}\,\|{}^{\{M\}}A_{N}\|^{-1}_{\mathbb{I}} = [\,{}^{m}A_{N}\,]_{m=1,\dots,M} \tag{2.54c} $$

Let us observe that ${}^{\{M\}}\hat{1}_{N;o}$, ${}^{\{M\}}A_{N}$ and ${}^{\{M\}}A_{N}$ are the representatives of the ${}^{M}\hat{y}_{N;o}$, ${}^{M}\varepsilon_{N}$ and ${}^{M}e_{N;o}$, respectively, in the isomorphism (2.47).

From the orthogonality conditions (2.54b), we obtain for $u=1,\dots,M$ and $(k_1,\dots,k_u) \in \mathrm{sym}\,{}^{u}L^{1}_{N-1}$

$$ ({}^{\{M\}}A_{N},\, {}^{\{M\}}1_{k_1\dots k_u})_{\mathbb{I}} = 0 \tag{2.55a} $$

while the norm of the error (2.54b) is given by

$$ ({}^{\{M\}}A_{N},\, {}^{\{M\}}1_{o})_{\mathbb{I}} = \|{}^{\{M\}}A_{N}\|_{\mathbb{I}} \tag{2.55b} $$

We notice that the 'normal equations' (2.55) are actually equivalent to (2.52), under the isomorphism (2.47).

2.3.7 Optimum generalized polynomial approximation

Given the space ${}^{\{M\}}\mathbf{Z}^{1}_{N-1}$ (2.44), let us consider

$$ {}^{\{M\}}\mathbf{Z}_N = \vee\{ {}^{M}z_o \, , \, {}^{\{M\}}Z^{1}_{N-1} \} \tag{2.56} $$

The N-th order estimate ${}^{M}\hat{z}_{N;o}$ of the element ${}^{M}z_o = [1\ 0\ \ldots\ 0]$ will be obtained as

$$ {}^{M}\hat{z}_{N;o} \triangleq {}^{\{M\}}P_{\mathbf{Z};N-1}^{1}\ {}^{M}z_o \ \in\ {}^{\{M\}}\mathbf{Z}^{1}_{N-1} \tag{2.57a} $$

where ${}^{\{M\}}P_{\mathbf{Z};N-1}^{1}$ denotes the orthogonal projection operator, taking projection on the subspace ${}^{\{M\}}\mathbf{Z}^{1}_{N-1}$. The N-th order approximation error will then be expressed as

$$ {}^{M}\Delta_{N;o} \triangleq {}^{\{M\}}P^{\perp\,1}_{\mathbf{Z};N-1}\ {}^{M}z_o = {}^{M}z_o - {}^{M}\hat{z}_{N;o} \ \perp\ {}^{\{M\}}\mathbf{Z}^{1}_{N-1} \tag{2.57b} $$

or, in a renormalized form,

$$ {}^{M}\mathbf{A}_N \triangleq {}^{M}\Delta_{N;o} \| {}^{M}\Delta_{N;o} \|^{-1}_{\mathbf{Z}} = [\, \mathbf{A}_N(z_1,\ldots,z_m) \,]_{m=1,\ldots,M} \tag{2.57c} $$

Let us observe that ${}^{M}\hat{z}_{N;o}$, ${}^{M}\Delta_{N;o}$ and ${}^{M}\mathbf{A}_N$ are the representatives of the ${}^{M}\hat{y}_{N;o}$, ${}^{M}\varepsilon_{N;o}$, ${}^{M}e_{N;o}$ and/or of the ${}^{\{M\}}\hat{1}_{N;o}$, ${}^{\{M\}}A_N$, ${}^{\{M\}}\mathbf{A}_N$ respectively, in the isomorphism (2.47).

The orthogonality conditions (2.57b) will imply for $u=1,\ldots,M$ and $(k_1,\ldots,k_u) \in \mathrm{sym}^{u}L^{1}_{N-1}$

$$ ({}^{M}\mathbf{A}_N \, , \, {}^{M}z_{k_1\ldots k_u})_{\mathbf{Z}} = 0 \tag{2.58a} $$

Moreover, we obtain

$$({}^{M}\mathbf{A}_N, {}^{M}z_o)_{\mathbf{Z}} = \|{}^{M}\Delta_{N;o}\|_{\mathbf{Z}} \tag{2.58b}$$

Let us observe that the 'normal equations' (2.58) are equivalent to the conditions (2.55),(2.52) and (2.31), under the isomorphism (2.47).

Finally, we can conclude that the nonlinear least-squares prediction filter problem can be equivalently stated in the spaces:

- of the regular Volterra functional polynomials;
- of the generalized coefficient-matrices;
- of the generalized z-polynomials.

This follows from the fact that the three spaces are isomorphic, as it is seen from (2.47). Therefore, the nonlinear stochastic estimation problem (i.e., the problem of optimum stochastic approximation of the output of the 'ideal' prediction filter) in the space of the Volterra functional polynomials ${}^{\{M\}}\mathbf{Y}_N$ can be equivalently considered as the problem of:

- the optimum deterministic approximation of the set of M-D impulse responses of the 'ideal' prediction filter in the space ${}^{\{M\}}\mathbf{I}_N$ of the generalized matrices (provided the higher-order covariance data of the underlying stochastic sequence are given);
- the optimum deterministic approximation of the set of M-D transfer functions of that filter, in the space ${}^{\{M\}}\mathbf{Z}_N$ of the generalized z-polynomials - in the AR case (provided the M-D spectral functions of the higher-order sequence are known).

3. GENERALIZED NONLINEAR LADDER-FILTERS

In the previous chapter we showed that the nonlinear least-squares prediction problem for higher-order stochastic sequences can be equivalently introduced algebraically and geometrically, in time- and in frequency-domains. Algebraic solution of the problem, yielding the nonlinear generalization of the Levinson algorithm, was proposed in Zarzycki and Dewilde (1983a,b); Zarzycki (1983); geometric derivation of the nonlinear prediction filter algorithm in the space of the Volterra functional polynomials was introduced in Zarzycki (1984a-e). Here we wish to present recursive solutions of the nonlinear least-squares prediction problem, using projection method in the spaces: of the generalized matrices, and of the generalized z-polynomials.

The nonlinear problem was stated in general case while its solution will be presented in case of the second-degree of the filter nonlinearity. This approach will show the most important features of the nonlinear treatment, avoiding unnecessary complexity of the derivations on one hand, on the other - the general nonlinear prediction filter algorithm will be obtained as a straightforward extension of the second-degree case. Since all quantities (i.e., subspaces, estimates and errors) considered in this chapter will be of the second-degree $(M=2)$, we shall drop from now on the superscript M in order to simplify notations.

We will show that the solution will result in a generalized (time-variant) nonlinear ladder-filter algorithm, being a natural generalization of the orthogonal linear ladder-filters.

3.1 Index-sets and their ordering

Let us consider for $n=0,\dots,N$ and $x=0,\dots,N-n$ the following index-sets (see also (2.16) with $m=1,2$)

$$ {}^1L_n^x = \{j_1\}_x^{x+n} = \{x,\dots,x+n\} \tag{3.1a} $$

$$ {}^2L_n^x = L_n^x \times L_n^x = \{j_1,j_2\}_{j_1,j_2=x}^{x+n} \tag{3.1b} $$

$$ \mathrm{sym}\,{}^2L_n^x = \{j_1,j_2\}_{j_1=x}^{x+n}\ {}_{j_2=j_1}^{x+n} $$

Then we can introduce the two types of the index-sets:

A) for $m=0,\dots,n+2$

$$ L_{n,m}^x \overset{\Delta}{=} \{x\} \cup L_{n-1}^{x+1} \cup \mathrm{sym}\,{}^2L_{n-1}^{x+1} $$

$$ \cup \{x+n+1\} \cup \{(x+n+1,x+n+1),\dots,(x+n+3-m,x+n+1)\} \tag{3.2a} $$

where x will denote a 'global' offset (backward shift from the reference point 0), n will stand for the 'global' order (number of previously completed columns) while m will indicate the 'local' order (number of elements in the last, and being completed, column);

B) for $v=0,\dots,n$ and $m=v+1,\dots,v+n+3$

$$ L_{n,m}^{x,v} \overset{\Delta}{=} \{x\} \cup \{(x,x),\dots,(x,x+v)\} \cup L_{n-1}^{x+1} \cup \mathrm{sym}\,{}^2L_{n-1}^{x+1} $$

$$ \cup \{x+n+1\} \cup \{(x+n+1,x+n+1),\dots,(x+n+4-m,x+n+1)\} \tag{3.2b} $$

where x will again denote the 'global' offset (in the 'uni-variate' subset, and in the top-row of the 'bi-variate' part of the $L_{n,m}^{x,v}$), v will indicate the last element in the top-row, n will be again the 'global' order while m will indicate the sum of the number of elements in the top-row and in the last column.

We can observe that

$$ {}^{1}L_{n}^{x} \cup \mathrm{sym}\,{}^{2}L_{n}^{x} = L_{n,n+1}^{x,n} \tag{3.2c} $$

and that the ordering is modulo $(n+2)$ as we have

$$ L_{n,m+n+2}^{x} = L_{n+1,m}^{x} \tag{3.3a} $$

$$ L_{n,m+n+2}^{x,v} = L_{n+1,m}^{x,v} \tag{3.3b} $$

Considering the element $\{x\}$ and the elements of the top-row, we introduce for further use the 'linear - forward' (L-forward), and 'bilinear - forward' (B-forward) index-sets of the n-th 'global' order, as follows

<u>L-forward</u>

$$ L_{n,o}^{x} = \{x\} \cup L_{n-1,n}^{x+1,n-1} \tag{3.4a} $$

<u>B-forward</u> (for $v=0,\dots,n$)

$$ L_{n,v+1}^{x,v} = \begin{cases} \{x,x\} \cup L_{n,o}^{x} & \text{if } v=0 \\ \{x,x+v\} \cup L_{n,v}^{x,v-1} & \text{if } v=1,\dots,n \end{cases} \tag{3.4b} $$

These index-sets will be applied later, in order to introduce sets of 'local' forward approximation errors.

Considering the element $\{x+n+1\}$ and the elements of the last column of the index-sets $L^{x,v}_{n,m}$, we define for further use the 'linear - backward (L-backward), and 'bilinear - backward' (B-backward) index-sets of the n-th 'global' order, as follows

<u>L-backward</u>

$$L^{x,n-1}_{n,o} = L^{x,n-1}_{n-1,n} \cup \{x+n\} \tag{3.5a}$$

<u>B-backward</u> (for m=1,...,n+1)

$$L^{x,n-1}_{n,m} = L^{x,n-1}_{n,m-1} \cup \{x+n+1-m, x+n\} \quad , \qquad m=1,\ldots,n \tag{3.5b}$$

$$L^{x,n}_{n,n+1} = L^{x,n-1}_{n,n} \cup \{x, x+n\} \quad , \qquad m=n+1 \tag{3.5c}$$

The backward index-sets will be used later, when sets of 'local' backward approximation errors will be considered.

Using (3.4), we can consider the following collection of the 'local' forward index-sets

$$\mathbb{L}^{x}_{f;n} = [\, L^{x}_{n,o} \quad L^{x,o}_{n,1} \quad \cdots \quad L^{x,n-1}_{n,n} \quad L^{x,n}_{n,n+1} \,] \tag{3.6a}$$

while the following collection of the 'local' backward index-sets results from (3.5)

$$\mathbb{L}^{x}_{b;n} = [\, L^{x,n-1}_{n,o} \quad L^{x,n-1}_{n,1} \quad \cdots \quad L^{x,n-1}_{n,n} \quad L^{x,n}_{n,n+1} \,] \tag{3.6b}$$

We can observe that the 'global' forward and backward index-sets (3.6) of the (n+1) order (i.e., $\mathbb{L}^{x}_{f;n+1}$ and $\mathbb{L}^{x}_{b;n+1}$) can be obtained in a recursive way, given the n-th order 'global' forward and backward index-sets $\mathbb{L}^{x}_{f;n}$ and $\mathbb{L}^{x}_{b;n}$. The higher-order 'global' index-sets will be derived via a 'global' order-update step $n \to n+1$, executed on the 'local' entries (3.4) and (3.5) of the sets (3.6), and described schematically in Fig.3.1.

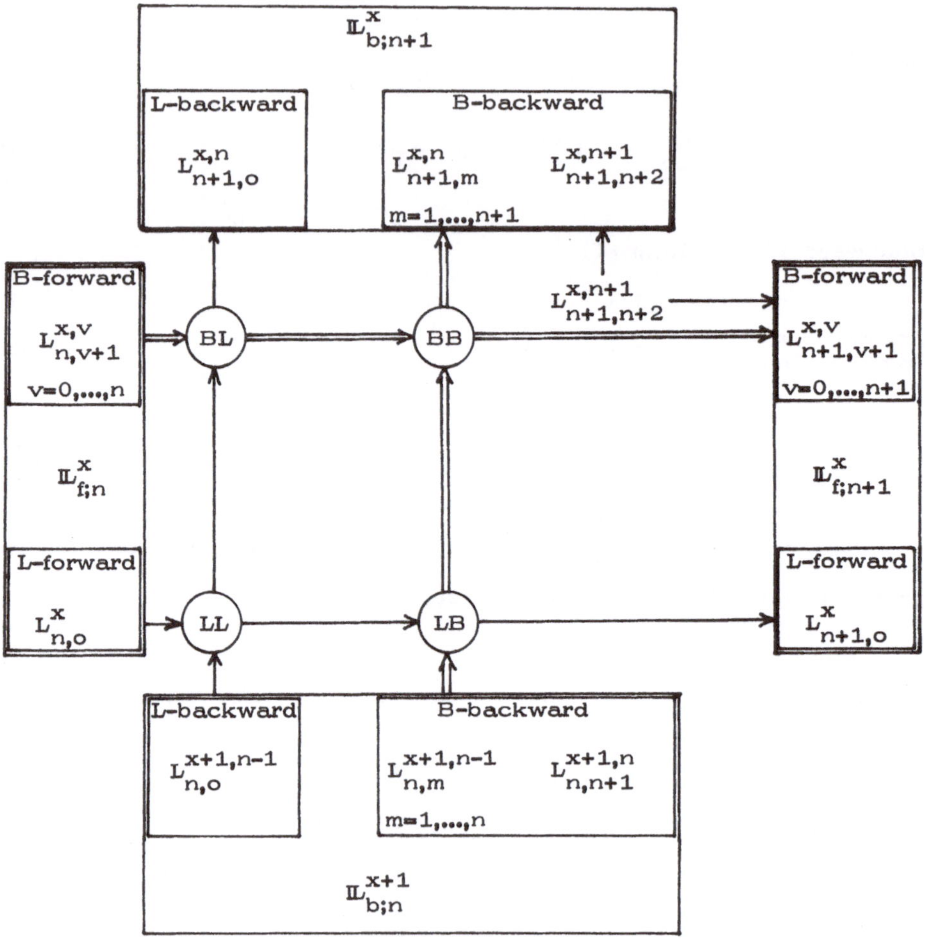

Fig.3.1 Order-update step $n \to n+1$ for the 'global' L-forward and L-backward index-sets.

The four types of the 'local' recursions, indicated in the diagram of Fig. 3.1 , can be interpreted as follows:

<u>LL - recursion</u>: the 'local' order-update for the 'uni-variate' parts of the L-forward and L-backward index-sets;

<u>LB - recursions:</u> the 'local' order-updates for the 'bi-variate' part of the L-forward index-set, and the 'local' order-update for the 'uni-variate' part of the B-backward index-sets;

<u>BL - recursions:</u> the 'local' order-update for the 'uni-variate' part of the B-forward index-sets, and the 'local' order-updates for the 'bi-variate' part of the L-backward index-set;

<u>BB - recursions:</u> the 'local' order-updates for the 'bi-variate' parts of the B-forward and B-backward index-sets.

For example, from (3.4a) and (3.5a) (with x replaced by $x+1$), it follows that the LL 'local' order-update index-set recursion can be expressed as

$$L^{x}_{n,1} = L^{x}_{n,o} \cup L^{x+1,n-1}_{n,o} = \underbrace{\{x\} \cup \rlap{\overbrace{\phantom{L^{x+1,n-1}_{n-1,n} \cup \{x+n+1\}}}^{L^{x+1,n-1}_{n,o}}} L^{x+1,n-1}_{n-1,n}}_{L^{x}_{n,o}} \cup \{x+n+1\} \qquad (3.7)$$

and can be schematically described as the following LL index-set section

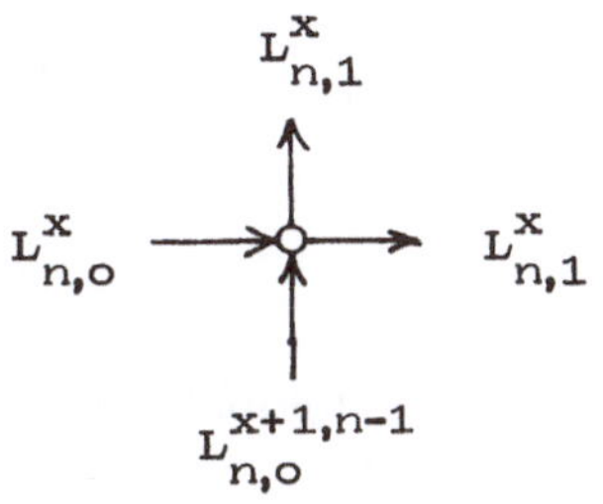

(3.8)

The remaining 'local' index-set recursions can be obtained in a similar way.

These recursions, together with the corresponding index-set section are presented in Appendix 1. We notice that the number of entries of the (n+1) 'global' order index-sets $\mathbb{L}^{x}_{f;n+1}$ and $\mathbb{L}^{x}_{b;n+1}$ is augmented by one in comparison with the n-th order sets $\mathbb{L}^{x}_{f;n}$ and $\mathbb{L}^{x}_{b;n}$, since the 'new' B-forward and B-backward index-set $L^{x,n+1}_{n+1,n+2}$ (associated with the 'right-upper' element $\{x,x+n+1\}$) is arriving to the scheme of Fig. 3.1 at the 'global' order-update step $n \rightarrow n+1$. The 'local' structure of this order-update step is presented in Fig. 3.2.

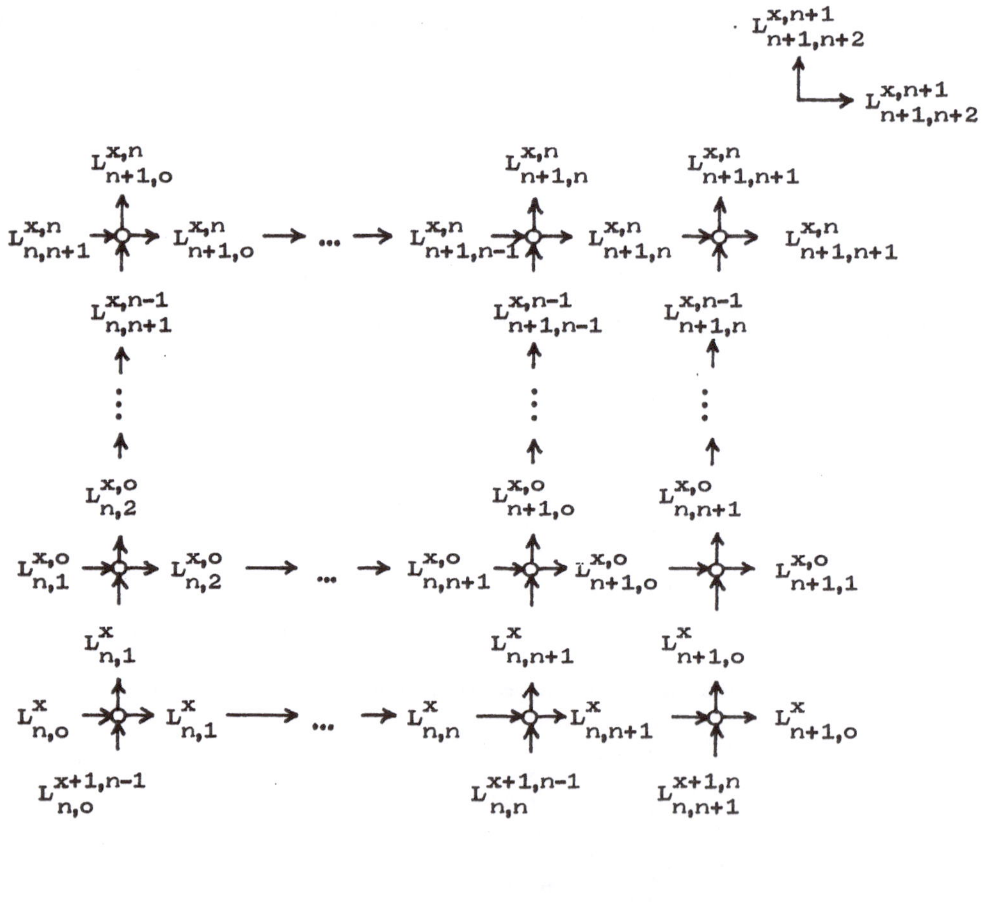

Fig. 3.2 'Local' structure of the 'global' order-update $n \rightarrow n+1$ index-set step.

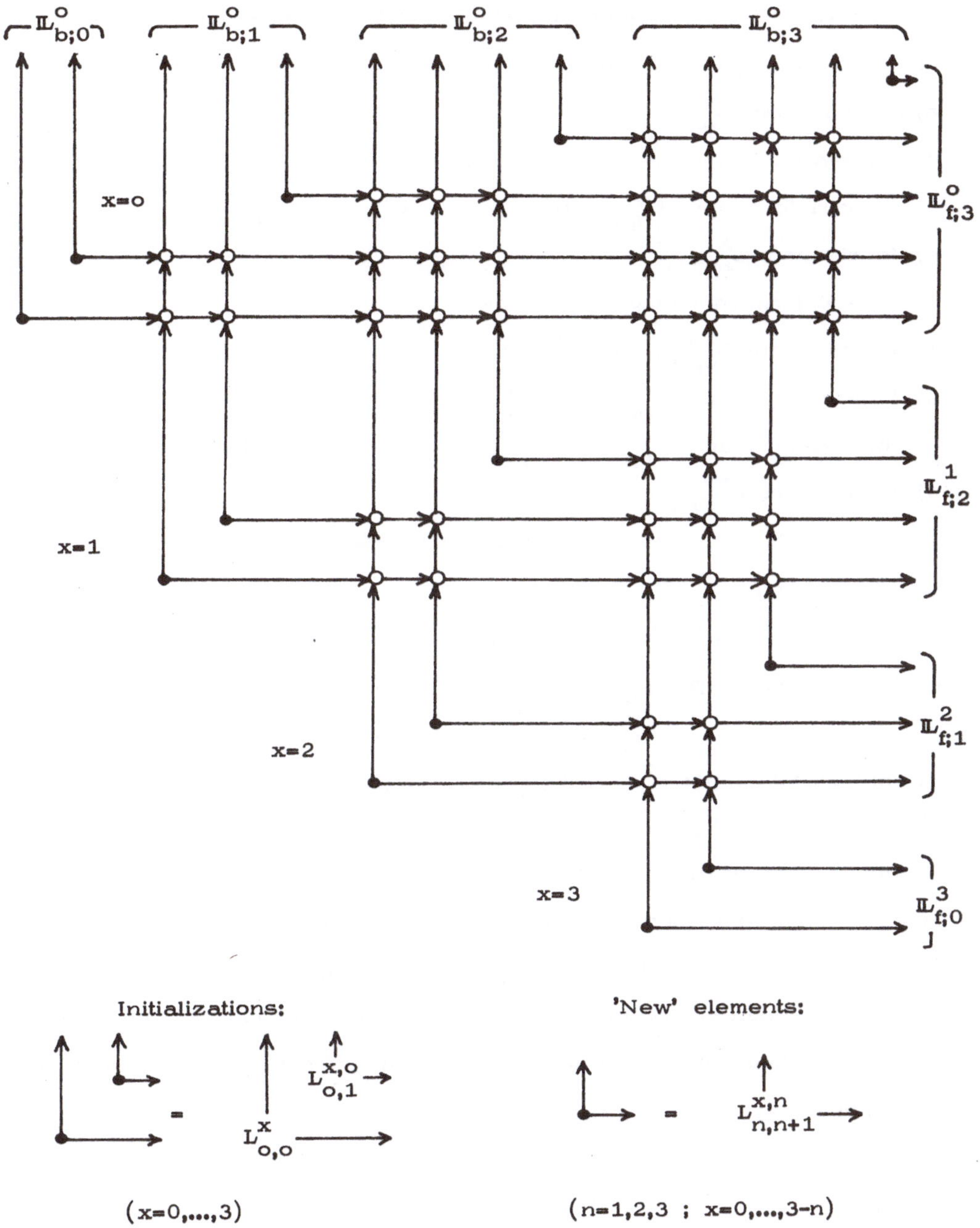

Fig. 3.3 'Local' structure of the third-order (N=3) index-set recursions. The symbols O indicate corresponding 'local' index-set recursions of Fig. 3.2.

From Fig. 3.2 it follows that the 'global' order-update recursion for the L-forward index-set will work as follows:

a) initialization: $L^{x}_{n,o}$

b) 'uni-variate' step: $L^{x}_{n,o} \rightarrow L^{x}_{n,1}$

c) 'bi-variate' steps: $L^{x}_{n,1} \rightarrow L^{x}_{n,2} \rightarrow \dots \rightarrow L^{x}_{n,n+1} \rightarrow L^{x}_{n+1,o}$

d) termination: $L^{x}_{n+1,o}$.

For the B-forward index-sets we get:

a) initialization: $L^{x,v}_{n,v+1}$ $(v=0,\dots,n)$

b) 'uni-variate' step: $L^{x,v}_{n,v+1} \rightarrow L^{x,v}_{n,v+2}$

c) 'bi-variate' steps: $L^{x,v}_{n,v+2} \rightarrow L^{x,v}_{n,v+3} \rightarrow \dots \rightarrow L^{x,v}_{n,n+1} \rightarrow L^{x,v}_{n+1,o} \rightarrow \dots \rightarrow L^{x,v}_{n+1,v+1}$

d) termination: $L^{x,v}_{n+1,v+1}$.

The L- and B-backward index-sets are updated as follows:

a) initializations: $L^{x+1,n-1}_{n,m}$ $(m=0,\dots,n)$

$L^{x+1,n}_{n,n+1}$ $(m=n+1)$

b) 'uni-variate' steps: $L^{x+1,n-1}_{n,m} \rightarrow L^{x}_{n,m+1}$ $(m=0,\dots,n)$

$L^{x+1,n}_{n,n+1} \rightarrow L^{x}_{n+1,o}$ $(m=n+1)$

c) 'bi-variate' steps: $L^{x}_{n,m+1} \rightarrow L^{x,o}_{n,m+2} \rightarrow L^{x,1}_{n,m+3} \rightarrow \dots \rightarrow L^{x,n-1}_{n+1,m-1} \rightarrow L^{x,n}_{n+1,m}$

d) termination: $L^{x,n}_{n+1,m}$.

We notice that each 'local' order-update step for the backward index-sets is associated with 'label-update' step. The 'local' structure of the third-order (N=3) index-set recursions is presented in Fig. 3.3.

The index-set recursions, derived in this section, will underly nonlinear ladder-filter algorithms presented in the subsequent paragraphs of this chapter.

3.2 Nonlinear filter algorithm: time-domain approach

In this paragraph we will derive a recursive solution to the second-degree nonlinear prediction problem, using projection method in the space of generalized coefficient-matrices, introduced in paragraph 2.3.2.

Let $\{y\}$ denote a fourth-order $(M=2)$ stochastic sequence, observed on the time-interval $[0,-1,\dots,-N]$ and represented by the random variables $y_0, y_{-1},\dots,y_{-N}$. Considering the index-sets $L^x_{n,m}$ (3.2a) and $L^{x,v}_{n,m}$ (3.2b), we define for $n=0,\dots,N$ and $x=0,\dots,N-n$ the following submatrices of the random variables and their products

$$Y^x_{n,m} = \begin{bmatrix} {}^1Y^x_{n,m} \\ \\ {}^2Y^x_{n,m} \end{bmatrix} = \begin{bmatrix} y_{-j_1} \\ \\ y_{-j_1} y_{-j_2} \end{bmatrix}_{(j_1,j_2) \in L^x_{n,m}} \tag{3.9}$$

and the submatrices $Y^{x,v}_{n,m}$, expressed by (3.9) with $L^x_{n,m}$ replaced by $L^{x,v}_{n,m}$. Then we can consider the following (2 2)-block, multi-indexed covariance submatrices $H^x_{n,m}$ and $H^{x,v}_{n,m}$, where the former is given by

$$H^x_{n,m} = \mathbb{E}\{Y^x_{n,m} \otimes \tilde{Y}^x_{n,m}\} = \begin{bmatrix} {}^{1\oplus 1}H^x_{n,m} & {}^{1\oplus 2}H^x_{n,m} \\ \\ {}^{2\oplus 1}H^x_{n,m} & {}^{2\oplus 2}H^x_{n,m} \end{bmatrix} =$$

$$= \begin{bmatrix} h_{j_1k_1} & h_{j_1k_1k_2} \\ \\ h_{j_1j_2k_1} & h_{j_1j_2k_1k_2} \end{bmatrix}_{(j_1,j_2,k_1,k_2) \in L^x_{n,m} \times L^x_{n,m}} \tag{3.10}$$

and the latter is expressed by (3.10) with $L^x_{n,m}$ replaced by $L^{x,v}_{n,m}$.

Now let

$$I^x_{n,m} = [\,{}^1I^x_{n,m} \quad {}^2I^x_{n,m}\,] = [\,1_{j_1} \quad 1_{j_1j_2}\,]_{(j_1,j_2) \in L^x_{n,m}} \tag{3.11a}$$

where for $(k_1,k_2) \in L^x_{n,m}$

$$1_{j_1} = [\,\delta_{k_1;j_1} \quad {}^2O\,] \quad ; \quad 1_{j_1j_2} = [\,{}^1O \quad \delta_{k_1k_2;j_1j_2}\,] \tag{3.11b}$$

with ${}^{1}O$ and ${}^{2}O$ being the one- and two-indexed zero-matrices, respectively, whose domains are the uni- and bi-variate parts of the index-set $L^{x}_{n,m}$, respectively. In a similar way we can introduce the matrix $I^{x,v}_{n,m}$.

Let us introduce the following, x-labeled and (x,v)-labeled, subspaces

$$\mathbf{I}^{x}_{n,m} = \vee\,\{I^{x}_{n,m}\} \quad ; \quad \mathbf{I}^{x,v}_{n,m} = \vee\,\{I^{x,v}_{n,m}\} \tag{3.12a}$$

We notice that $\mathbf{I}^{1,N-1}_{N-1,N}$ will be precisely the space ${}^{\{2\}}\mathbf{I}^{1}_{N-1}$ (expressed by (2.37) with M=2) while the 'biggest' space will be $\mathbf{I}^{o,N}_{N,N+1}$. Each element $F^{x}_{n,m}$ of the subspace $\mathbf{I}^{x}_{n,m}$ will be expressed as

$$F^{x}_{n,m} = [\, f^{x}_{n,m;j_1-x} \quad f^{x}_{n,m;j_1-x,j_2-x} \,]_{(j_1,j_2)\,\epsilon\, L^{x}_{n,m}} \tag{3.12b}$$

i.e., $F^{x}_{n,m}$ will be a two-block (row), multi-indexed coefficient-matrix whose domain is $DF^{x}_{n,m} = L^{x}_{n,m}$. Similarly, each element of the subspace $\mathbf{I}^{x,v}_{n,m}$ will be $F^{x,v}_{n,m}$ with domain $DF^{x,v}_{n,m} = L^{x,v}_{n,m}$. Following (2.39a), we introduce a family of inner-products on the subspaces (3.12a) as

$$(F^{x}_{n,m}, G^{x}_{n,m})_{\mathbf{I}^{x}_{n,m}} = F^{x}_{n,m}\cdot H^{x}_{n,m}\cdot \tilde{G}^{x}_{n,m} \tag{3.13a}$$

$$(F^{x,v}_{n,m}, G^{x,v}_{n,m})_{\mathbf{I}^{x,v}_{n,m}} = F^{x,v}_{n,m}\cdot H^{x,v}_{n,m}\cdot \tilde{G}^{x,v}_{n,m} \tag{3.13b}$$

According to (3.4) and (3.5), we can introduce for $n=0,\dots,N$ and $x=0,\dots,N-n$ the subsequent subspaces of the $\mathbf{I}^{o,N}_{N,N+1}$

<u>L-forward</u>

$$\mathbf{I}^{x}_{n,o} = \vee\,\{I^{x}_{n,o}\} \tag{3.14a}$$

B-forward (for $v=0,\dots,n$)

$$\mathbb{I}^{x,v}_{n,v+1} = v\{I^{x,v}_{n,v+1}\} \tag{3.14b}$$

L-backward

$$\mathbb{I}^{x,n-1}_{n,o} = v\{I^{x,n-1}_{n,o}\} \tag{3.14c}$$

B-backward (for $m=1,\dots,n+1$)

$$\mathbb{I}^{x,n-1}_{n,m} = v\{I^{x,n-1}_{n,m}\} \quad , \qquad m=1,\dots,n \tag{3.14d}$$

$$\mathbb{I}^{x,n}_{n,n+1} = v\{I^{x,n}_{n,n+1}\} \quad , \qquad m=n+1 \tag{3.14e}$$

3.2.1 'Local' estimates and errors

Denoting by $P^{x}_{\mathbb{I};n,m}$ $(P^{x,v}_{\mathbb{I};n,m})$ the orthogonal projection operators, taking projection on the subspaces $\mathbb{I}^{x}_{n,m}$ $(\mathbb{I}^{x,v}_{n,m})$, we will introduce the subsequent 'local' estimates and errors as follows:

L-forward

Following (3.4a),(3.11),(3.12) and (3.14a), we can rewrite $\mathbb{I}^{x}_{n,o}$ as

$$\mathbb{I}^{x}_{n,o} = v\{1_x \ , \ I^{x+1,n-1}_{n-1,n}\} \tag{3.15a}$$

We define the L-forward n-th order estimate $\hat{1}^{x}_{n,o}$ of the 1_x as

$$\hat{1}^{x}_{n,o} \overset{\Delta}{=} P^{\;x+1,n-1}_{\mathbb{I};n-1,n}\, 1_x \in \mathbb{I}^{x+1,n-1}_{n-1,n} \tag{3.15b}$$

Let 0_{j_1} denote the zero-entry with 'coordinate' j_1. Then the L-forward n-th order approximation error, corresponding to the estimate $\hat{1}^x_{n,o}$, will be expressed as

$$A^x_{n,o} \triangleq P^{\perp\, x+1,n-1}_{\mathbb{I};n-1,n} \, 1_x = 1_x - [0_o \quad \hat{1}^x_{n,o}] \perp \mathbb{I}^{x+1,n-1}_{n-1,n} \tag{3.16a}$$

(since the estimate $\hat{1}^x_{n,o}$ is considered here as an element of the subspace $\mathbb{I}^x_{n,o}$). This error can be rewritten in a renormalized form

$$\mathrm{A}^x_{n,o} = A^x_{n,o} \| A^x_{n,o} \|^{-1}_{\mathbb{I}^x_{n,o}} =$$

$$= [a^x_{n,o;j_1-x} \quad a^x_{n,o;j_1-x,j_2-x}]_{(j_1,j_2) \in L^x_{n,o}} \tag{3.16b}$$

in accordance with (3.12b). We can observe that the estimate (3.15b) and the errors (3.16) are precisely the ${}^{\{M\}}1_{N;o}$, ${}^{\{M\}}A_N$ and ${}^{\{M\}}\mathrm{A}_N$ (2.54) respectively, if $M=2$, $x=0$ and $n=N$.

<u>B-forward</u> (for $v=0,\ldots,n$)

Using (3.4b), we can rewrite the subspaces $\mathbb{I}^{x,v}_{n,v+1}$ (3.14b) as

$$\mathbb{I}^{x,v}_{n,v+1} = \begin{cases} \vee \{ 1_{x,x} \,,\, \mathrm{I}^x_{n,o} \} & \text{if } v=0 \\ \vee \{ 1_{x,x+v} \,,\, \mathrm{I}^{x,v-1}_{n,v} \} & \text{if } v=1,\ldots,n \end{cases} \tag{3.17a}$$

Then we can introduce the B-forward estimates

$$1^{x,v}_{n,v+1} \triangleq \begin{cases} P^x_{\mathbb{I};n,o} \, 1_{x,x} \in \mathbb{I}^x_{n,o} & \text{if } v=0 \\ P^{x,v-1}_{\mathbb{I};n,v} \, 1_{x,x+v} \in \mathbb{I}^{x,v-1}_{n,v} & \text{if } v=1,\ldots,n \end{cases} \tag{3.17b}$$

Let $0_{j_1,j_2}$ stand for the zero-entry with 'coordinates' (j_1,j_2). Then the B-forward approximation errors, corresponding to the estimates (3.17b), will be expressed as

$$A^{x,v}_{n,v+1} \triangleq \begin{cases} P^{\perp\ x}_{\mathbb{I};n,o}\ 1_{x,x} \perp \mathbb{I}^{x}_{n,o} & \text{if}\quad v=0 \\ P^{\perp\ x,v-1}_{\mathbb{I};n,v}\ 1_{x,x+v} \perp \mathbb{I}^{x,v-1}_{n,v} & \text{if}\quad v=1,\ldots,n \end{cases} \qquad (3.18a)$$

i.e.,

$$A^{x,v}_{n,v+1} = 1_{x,x+v} - [\,0_{o,v}\quad \hat{1}^{x,v}_{n,v+1}\,] \qquad (3.18b)$$

(as the estimates $\hat{1}^{x,v}_{n,v+1}$ are treated here as elements of $\mathbb{I}^{x,v}_{n,v+1}$). The B-forward errors (3.18) can be expressed in the renormalized form as

$$\underline{A}^{x,v}_{n,v+1} = A^{x,v}_{n,v+1}\,\|A^{x,v}_{n,v+1}\|^{-1}_{\mathbb{I}^{x,v}_{n,v+1}} =$$

$$= [\,a^{x,v}_{n,v+1;j_1-x}\quad a^{x,v}_{n,v+1;j_1-x,j_2-x}\,] \quad (j_1,j_2) \in L^{x,v}_{n,v+1} \qquad (3.18c)$$

L-backward

Let us rewrite $\mathbb{I}^{x,n-1}_{n,o}$ (3.14c), using (3.5a), as

$$\mathbb{I}^{x,n-1}_{n,o} = \vee\ \{\mathbb{I}^{x,n-1}_{n-1,n}\ ,\ 1_{x+n}\} \qquad (3.19a)$$

so that we will define the L-backward estimate as

$$\check{1}^{x,n-1}_{n,o} \triangleq P^{\ x,n-1}_{\mathbb{I};n-1,n}\ 1_{x+n} \in \mathbb{I}^{x,n-1}_{n-1,n} \qquad (3.19b)$$

The L-backward approximation error will then be

$$B^{x,n-1}_{n,o} \triangleq P^{\perp\ x,n-1}_{\mathbb{I};n-1,n}\ 1_{x+n} = 1_{x+n} - [\,\check{1}^{x,n-1}_{n,o}\quad 0_n\,] \perp \mathbb{I}^{x,n-1}_{n-1,n} \qquad (3.20a)$$

or, in a renormalized form,

$$\mathrm{B}_{n,o}^{x,n-1} = \mathcal{B}_{n,o}^{x,n-1} \| \mathcal{B}_{n,o}^{x,n-1} \|_{\mathbb{I}_{n,o}^{x,n-1}}^{-1} =$$

$$= [b_{n,o;x+n-j_1}^{x,n-1} \quad b_{n,o;x+n-j_1,x+n-j_2}^{x,n-1}]_{(j_1,j_2) \in L_{n,o}^{x,n-1}} \qquad (3.20b)$$

<u>B-backward</u> (for $m=1,\ldots,n+1$)

Following (3.5) and (3.14), we can write

$$\mathbb{I}_{n,m}^{x,n-1} = \vee \{ \mathbb{I}_{n,m-1}^{x,n-1} \ , \ 1_{x+n+1-m,x+n} \} \ , \quad m=1,\ldots,n \qquad (3.21a)$$

$$\mathbb{I}_{n,n+1}^{x,n} = \vee \{ \mathbb{I}_{n,n}^{x,n-1} \ , \ 1_{x,x+n} \} \ , \quad m=n+1 \qquad (3.21b)$$

Hence, the B-backward estimates will be expressed as

$$\check{1}_{n,m}^{x,n-1} \overset{\Delta}{=} P_{\mathbb{I};n,m-1}^{x,n-1} \, 1_{x+n+1-m,x+n} \in \mathbb{I}_{n,m-1}^{x,n-1} \ , \quad m=1,\ldots,n \qquad (3.21c)$$

$$\check{1}_{n,n+1}^{x,n} \overset{\Delta}{=} P_{\mathbb{I};n,n}^{x,n-1} \, 1_{x,x+n} \in \mathbb{I}_{n,n}^{x,n-1} \ , \quad m=n+1 \qquad (3.21d)$$

Consequently, the B-backward approximation errors will be defined as

$$\mathcal{B}_{n,m}^{x,n-1} \overset{\Delta}{=} P_{\mathbb{I};n,m-1}^{\perp\, x,n-1} \, 1_{x+n+1-m,x+n} \perp \mathbb{I}_{n,m-1}^{x,n-1} \ , \quad m=1,\ldots,n \qquad (3.22a)$$

$$\mathcal{B}_{n,n+1}^{x,n} \overset{\Delta}{=} P_{\mathbb{I};n,n}^{\perp\, x,n-1} \, 1_{x,x+n} \perp \mathbb{I}_{n,n}^{x,n-1} \ , \quad m=n+1 \qquad (3.22b)$$

so that

$$\mathcal{B}_{n,m}^{x,n-1} = 1_{x+n+1-m,x+n} - [\check{1}_{n,m}^{x,n-1} \quad 0_{n+1-m,n}] \qquad (3.22c)$$

$$\mathcal{B}_{n,n+1}^{x,n} = 1_{x,x+n} - [\check{1}_{n,n+1}^{x,n} \quad 0_{o,n}] \qquad (3.22d)$$

The B-backward errors can be expressed in a renormalized form

$$B^{x,n-1}_{n,m} = \mathcal{B}^{x,n-1}_{n,m} \| \mathcal{B}^{x,n-1}_{n,m} \|^{-1}_{\mathbb{I}^{x,n-1}_{n,m}} , \qquad m=1,\dots,n \tag{3.22e}$$

$$B^{x,n}_{n,n+1} = \mathcal{B}^{x,n}_{n,n+1} \| \mathcal{B}^{x,n}_{n,n+1} \|^{-1}_{\mathbb{I}^{x,n}_{n,n+1}} , \qquad m=n+1 \tag{3.22f}$$

similarly as the L-backward error $B^{x,n-1}_{n,o}$ (3.20b).

3.2.2 Decomposition of subspaces

Following the considerations of the previous paragraph, we can consider the 'local' decompositions of subspaces:

L-forward

Since $A^{x}_{n,o}$ is in $\mathbb{I}^{x}_{n,o}$, but is orthogonal to $\mathbb{I}^{x+1,n-1}_{n-1,n}$, as it follows from (3.16), we can write

$$\mathbb{I}^{x}_{n,o} = \mathbb{I}^{x+1,n-1}_{n-1,n} \oplus \vee\{A^{x}_{n,o}\} \tag{3.23a}$$

which implies the 'local' decomposition of projection operators

$$P^{x}_{\mathbb{I};n,o} = P^{x+1,n-1}_{\mathbb{I};n-1,n} + P^{x}_{A;n,o} \tag{3.23b}$$

where $P^{x}_{A;n,o}$ denotes the orthogonal projection operator, taking projection on the span of $A^{x}_{n,o}$.

B-forward (for $v=0,\dots,n$)

Since $A^{x,v}_{n,v+1}$ belongs to $\mathbb{I}^{x,v}_{n,v+1}$ but is orthogonal to: $\mathbb{I}^{x}_{n,o}$ (if $v=0$), and to $\mathbb{I}^{x,v-1}_{n,v}$ (if $v=1,\dots,n$), in accordance with (3.18), we obtain

$$\mathbb{I}^{x,v}_{n,v+1} = \begin{cases} \mathbb{I}^{x}_{n,o} \oplus \vee\{A^{x,o}_{n,1}\} , & v=0 \\ \mathbb{I}^{x,v-1}_{n,v} \oplus \vee\{A^{x,v}_{n,v+1}\} , & v=1,\dots,n \end{cases} \tag{3.24a}$$

This implies

$$P^{x,v}_{\mathbb{I};n,v+1} = \begin{cases} P^{x}_{\mathbb{I};n,o} + P^{x,o}_{A;n,o} , & v=0 \\ P^{x,v-1}_{\mathbb{I};n,v} + P^{x,v}_{A;n,v+1} , & v=1,\dots,n \end{cases} \tag{3.24b}$$

where $P^{x,v}_{A;n,v+1}$ is the projection operator on the subspace spanned by $A^{x,v}_{n,v+1}$.

L-backward

Observing that $B^{x,n-1}_{n,o}$ belongs to the subspace $\mathbb{I}^{x,n-1}_{n,o}$ but is orthogonal to $\mathbb{I}^{x,n-1}_{n-1,n}$, see (3.20), we can write

$$\mathbb{I}^{x,n-1}_{n,o} = \mathbb{I}^{x,n-1}_{n-1,n} \oplus \vee\{B^{x,n-1}_{n,o}\} \tag{3.25a}$$

resulting in the decomposition

$$P^{x,n-1}_{\mathbb{I};n,o} = P^{x,n-1}_{\mathbb{I};n-1,n} + P^{x,n-1}_{B;n,o} \tag{3.25b}$$

where $P^{x,n-1}_{B;n,o}$ is the projection operator on the span of $B^{x,n-1}_{n,o}$.

B-backward (for $m=1,\dots,n+1$)

From (3.22) it follows that

$$\mathbb{I}^{x,n-1}_{n,m} = \mathbb{I}^{x,n-1}_{n,m-1} \oplus \vee\{B^{x,n-1}_{n,m}\} , \quad m=1,\dots,n \tag{3.26a}$$

$$\mathbb{I}^{x,n}_{n,n+1} = \mathbb{I}^{x,n-1}_{n,n} \oplus \vee\{B^{x,n}_{n,n+1}\} \quad , \quad m=n+1 \qquad (3.26b)$$

This implies

$$P^{x,n-1}_{\mathbb{I};n,m} = P^{x,n-1}_{\mathbb{I};n,m-1} + P^{x,n-1}_{B;n,m} \quad , \quad m=1,\ldots,n \qquad (3.26c)$$

$$P^{x,n}_{\mathbb{I};n,n+1} = P^{x,n-1}_{\mathbb{I};n,n} + P^{x,n}_{B;n,n+1} \quad , \quad m=n+1 \qquad (3.26d)$$

where $P^{x,n-1}_{B;n,m}$ $(P^{x,n}_{B;n,n+1})$ is the projection operator on the subspace spanned by $B^{x,n-1}_{n,m}$ $(B^{x,n}_{n,n+1})$.

3.2.3 Orthonormal bases

We can observe that the L- and B-forward errors $A^{x}_{n,o}$ and the $A^{x,v}_{n,v+1}$ $(v=0,\ldots,n)$ form the ON set in the space of generalized matrices. In order to show that, let us consider two B-forward errors $A^{x,v}_{n,v+1}$ and $A^{x,u}_{n,u+1}$, and let us assume that $v < u$. Since $A^{x,u}_{n,u+1} \perp \mathbb{I}^{x,u-1}_{n,u} \supset \mathbb{I}^{x,v}_{n,v+1}$, and $A^{x,v}_{n,v+1}$ belongs to $\mathbb{I}^{x,v}_{n,v+1}$, we obtain for $v=0,\ldots,n$

$$A^{x,v}_{n,v+1} \perp A^{x,u}_{n,u+1} \qquad (3.27a)$$

In a similar way we can show that for $v=0,\ldots,n$

$$A^{x,v}_{n,v+1} \perp A^{x}_{n,o} \qquad (3.27b)$$

Consequently, the entries of

$$\mathbb{A}^{x}_{n} = [\,A^{x}_{n,o} \quad A^{x,o}_{n,o} \quad \cdots \quad A^{x,n-1}_{n,n} \quad A^{x,n}_{n,n+1}] \qquad (3.28)$$

will form the ON set.

If we introduce, according to (3.15a) and (3.17a), the following set

$$[1_x \quad 1_{x,x} \quad \cdots \quad 1_{x,x+n-1} \quad 1_{x,x+n}] \tag{3.29}$$

then (3.28) will be the orthonormalized version of that set. Using (3.28) we can write the following 'global' orthogonal decomposition of subspaces

$$\mathbb{I}^{x,n}_{n,n+1} = \mathbb{I}^{x+1,n-1}_{n-1,n} \oplus v\{\mathbf{A}^x_n\} \tag{3.30a}$$

where

$$v\{\mathbf{A}^x_n\} = v\{A^x_{n,o}\} \oplus \sum_{v=0}^{n} \oplus\, v\{A^{x,v}_{n,v+1}\} \tag{3.30b}$$

The decomposition (3.30) will result in the 'global' decomposition of projection operators

$$P^{x,n}_{\mathbb{I};n,n+1} = P^{x+1,n-1}_{\mathbb{I};n-1,n} + P^x_{\mathbf{A};n} \tag{3.30c}$$

with

$$P^x_{\mathbf{A};n} = P^x_{A;n,o} + \sum_{v=0}^{n} P^{x,v}_{A;n,v+1} \tag{3.30d}$$

This result implies the following 'global' orthogonal decomposition of the 'biggest' subspace

$$\mathbb{I}^{o,N}_{N,N+1} = \sum_{x=0}^{N} \oplus\, v\{\mathbf{A}^x_{N-x}\} \tag{3.31}$$

Consequently, $\mathbf{A}^x_n$ (3.28) can be interpreted as the 'global' ON basis of the space of the two-block, multi-indexed coefficient-matrices. The entries of (3.28) can therefore be treated as the 'local' ON basis of that space. We shall call $\mathbf{A}^x_n$ the 'forward' ON basis of the space of generalized matrices.

In a similar way we can show that the L- and B-backward errors $B^{x,n-1}_{n,m}$ $(m=0,\ldots,n)$ and $B^{x,n}_{n,n+1}$ will form another ON set in the space of the generalized matrices. Considering two B-backward errors $B^{x,n-1}_{n,m}$ and $B^{x,n-1}_{n,w}$ $(m,w=1,\ldots,n)$, and assuming $m \quad w$, we can observe that $B^{x,n-1}_{n,w} \perp \mathbb{I}^{x,n-1}_{n,w-1} \supset \mathbb{I}^{x,n-1}_{n,m} \ni B^{x,n-1}_{n,m}$. Thus, we get for $m,w=1,\ldots,n$

$$B^{x,n-1}_{n,m} \perp B^{x,n-1}_{n,w} \tag{3.32a}$$

Similarly, we can show that (3.32a) holds for $m,w=0,1,\ldots,n$, and moreover, that for $m=0,\ldots,n$

$$B^{x,n-1}_{n,m} \perp B^{x,n}_{n,n+1} \tag{3.32b}$$

Therefore, the entries of

$$\mathbb{B}^{x}_{n} = [\, B^{x,n-1}_{n,o} \quad B^{x,n-1}_{n,1} \quad \cdots \quad B^{x,n-1}_{n,n} \quad B^{x,n}_{n,n+1} \,] \tag{3.33}$$

will form the ON set. Considering (due to (3.19) and (3.21)) the set

$$[\, 1_{x+n} \quad 1_{x+n,x+n} \quad \cdots \quad 1_{x+1,x+n} \quad 1_{x,x+n} \,] \tag{3.34}$$

we notice that (3.33) is the orthonormalized version of (3.34). Consequently, we obtain the following 'global' orthogonal decomposition of subspaces

$$\mathbb{I}^{x,n}_{n,n+1} = \mathbb{I}^{x,n-1}_{n-1,n} \oplus \vee\{\mathbb{B}^{x}_{n}\} \tag{3.35a}$$

where

$$\vee\{\mathbb{B}^{x}_{n}\} = \sum_{m=0}^{n} \oplus \vee\{B^{x,n-1}_{n,m}\} \oplus \vee\{B^{x,n}_{n,n+1}\} \tag{3.35b}$$

The decompositions (3.35) imply the 'global' decomposition of the projection operators

$$P_{\mathbb{I};n,n+1}^{x,n} = P_{\mathbb{I};n-1,n}^{x,n-1} + P_{\mathbb{B};n}^{x} \qquad (3.35c)$$

where

$$P_{\mathbb{B};n}^{x} = \sum_{m=0}^{n} P_{B;n,m}^{x,n-1} + P_{B;n,n+1}^{x,n} \qquad (3.35d)$$

Hence, we obtain the second kind of the 'global' decomposition of the 'biggest' subspace, in the form

$$\mathbb{I}_{N,N+1}^{o,N} = \sum_{n=0}^{N} \oplus \ v\,\{\mathbb{B}_n^o\} \qquad (3.36)$$

We shall call $\mathbb{B}_n^x$ the global 'backward' ON basis of the space of the generalized matrices. The entries (3.33) of $\mathbb{B}_n^x$ will therefore be termed the 'local' backward ON basis of that space.

3.2.4 Generalized Cholesky factorizations

Let us introduce the following multi-indexed 'upper-triangular' matrix

$$A_N = \mathrm{col}\,[\,\mathbb{A}_{N-x}^{x}\,]_{x=0,\dots,N} = \mathrm{col}\,[\mathbb{A}_N^o \ \dots \ \mathbb{A}_n^x \ \dots \ \mathbb{A}_o^N\,] \qquad (3.37a)$$

and let

$$B_N = \mathrm{col}\,[\,\mathbb{B}_n^o\,]_{n=0,\dots,N} = \mathrm{col}\,[\mathbb{B}_o^o \ \dots \ \mathbb{B}_n^o \ \dots \ \mathbb{B}_N^o\,] \qquad (3.37b)$$

be the corresponding multi-indexed 'lower-triangular' matrix. Recalling the inner-product, introduced in the 'biggest' subspace of generalized matrices and expressed by (3.13b) with $x=0$, $v=N$, $n=N$, $m=N+1$), we can observe that

$$(A_N , A_N)_{\mathbb{I}^{o,N}_{N,N+1}} = A_N \cdot H^{o,N}_{N,N+1} \cdot \tilde{A}_N = \mathbf{1} \qquad (3.38a)$$

$$(B_N , B_N)_{\mathbb{I}^{o,N}_{N,N+1}} = B_N \cdot H^{o,N}_{N,N+1} \cdot \tilde{B}_N = \mathbf{1} \qquad (3.38b)$$

where $\mathbf{1}$ is the unit (2×2)-block, multi-indexed matrix (i.e., the unit-entries of the $\mathbf{1}$ will have coordinates (j_1,j_1) and (j_1,j_1,j_2,j_2)) whose domain equals $L^{o,N}_{N,N+1} \times L^{o,N}_{N,N+1}$. The relations (3.38a) can be shown by writing the matrices A_N and $H^{o,N}_{N,N+1}$ 'on the paper' and then, observing that the LHS of (3.38a) is a symmetric matrix while the RHS is upper-triangular (with unit diagonal-entries) so that the RHS is the unit multi-indexed matrix $\mathbf{1}$. Equation (3.38b) can be shown in a similar way.

Consequently, the ON conditions (3.38) express the 'forward', resp. 'backward' generalized Cholesky factorization of the block, multi-indexed covariance matrix of the fourth-order stochastic sequence. We can observe that (3.38) can be generalized to the 2M-th order sequences in a straightforward way. We also remark that, assuming that the underlying sequence is of second-order (and, hence, is represented by the two-indexed left-upper block-entry of $H^{o,N}_{N,N+1}$), conditions (3.38) will reduce to the Cholesky factorization of the two-indexed covariance matrix associated with the linear least-squares prediction problem (see Deprettere and Lie (1980)).

3.2.5 M-D Fourier series expansion

We can observe that the 'biggest' subspace $\mathbb{I}^{1,N-1}_{N-1,N}$ will correspond to the biggest 'estimation space' $\mathbf{Y}^{1,N-1}_{N-1,N}$, under the isomorphism (2.47). The latter space has been considered in order to obtain the generalized ON expansion (in the form of the stochastic functional M-D Fourier series) for the random variable y_o , in the space of the Volterra functional polynomials, yielding the optimum solution to the (second-degree) nonlinear least-squares prediction problem (see e.g., Zarzycki (1984b)). We notice that the 1_o is actually the representative of y_o in the isomorphism (2.47) so that the subspace $\mathbb{I}^{1,N-1}_{N-1,N}$ will play the role of the mentioned above 'estimation space'; i.e., it will be used in order to obtain the desired ON development of the element 1_o . Following (3.35), we get

$$\mathbb{I}^{1,N-1}_{N-1,N} = \sum_{n=0}^{N-1} \oplus \vee \{\mathbb{B}^1_n\} \tag{3.39a}$$

so that

$$P_{\mathbb{I};N-1,N}^{\;1,N-1} = \sum_{n=0}^{N-1} P_{\mathbb{B};n}^{\;1} \tag{3.39b}$$

This decomposition can be used in order to obtain the M-D Fourier series expansion for the 1_o , in the form

$$1_o \sim \sum_{n=0}^{N-1} (1_o, \mathbb{B}^1_n)\, \mathbb{B}^1_n \tag{3.39c}$$

We shall see later that (3.39) will imply a generalized Fourier expansion for the set of M-D impulse responses of the (second-degree) nonlinear filter, yielding the optimum solution to the nonlinear least-squares prediction problem. In other words, we will show that the nonlinear problem is solved once the 'backward' ON basis of the space of generalized matrices is obtained, and that the basis can be computed in a recursive way, resulting in the nonlinear ladder-filter algorithm.

3.2.6 Order-update recursions

We wish to show that the 'global' forward as well as backward ON bases can be introduced recursively, via a set of appropriately defined 'local' and 'global' order-update recursions. In other words, we wish to show how the (n+1) 'global' order, forward $\mathbb{A}^x_{n+1}$ and backward $\mathbb{B}^x_{n+1}$, solutions can be obtained recursively from the n-th 'global' order solutions $\mathbb{A}^x_n$ and $\mathbb{B}^{x+1}_n$. Let us observe that the (mutually orthonormal) entries of $\mathbb{A}^x_{n+1}$ ($\mathbb{B}^x_{n+1}$) should be all orthogonal to the subspace $\mathbb{I}^{x+1,n}_{n,n+1}$ ($\mathbb{I}^{x,n}_{n,n+1}$). Using the decompositions (3.35) and (3.30), we will show that $\mathbb{A}^x_{n+1}$ and $\mathbb{B}^x_{n+1}$ can be obtained via a set of appropriately introduced 'local' order-update recursions, executed on the L- and B-, forward and backward entries of the $\mathbb{A}^x_n$ and $\mathbb{B}^{x+1}_n$. These recursions (based on the index-set recursions, considered in paragraph 3.3) will preserve mutual orthonormality of the 'local' entries of the 'global' forward and backward approximation errors after each 'local' order-update step. Consequently, the entries of the $\mathbb{A}^x_{n+1}$ and $\mathbb{B}^x_{n+1}$, obtained that way, will form the ON sets again.

Reinterpreting the index-set scheme of Fig. 3.1, the desired 'global' order-update step can be described as follows

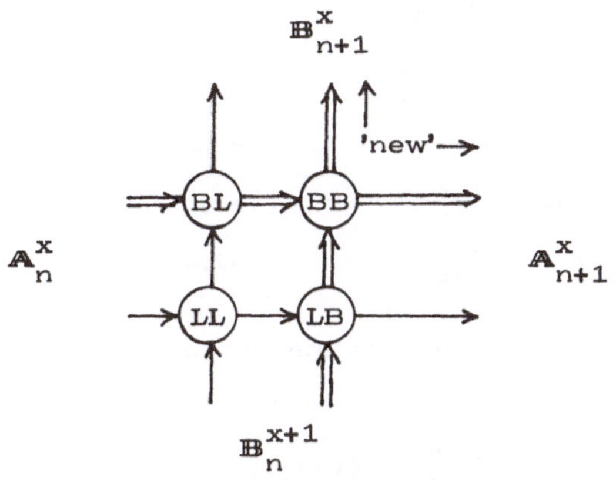

(3.40)

Considering (3.28) and (3.33) with n replaced by $n+1$, we notice that the number of entries of $\mathbb{A}^x_n$ and of $\mathbb{B}^x_n$ will be augmented by one, during the order-update step $n \to n+1$, yielding $\mathbb{A}^x_{n+1}$ and $\mathbb{B}^x_{n+1}$

(much like in the index-set scheme of Fig. 3.1). Thus, each 'global' order-update step will introduce the following 'new' B-forward and B-backward approximation error

$$A^{x,n+1}_{n+1,n+2} = B^{x,n+1}_{n+1,n+2} \qquad (3.41a)$$

whose unnormalized version will be expressed as

$$\mathcal{A}^{x,n+1}_{n+1,n+2} = \mathcal{B}^{x,n+1}_{n+1,n+2} = P^{\perp\ x,n}_{\mathbb{I}\ ;n+1,n+1}\ {}^{1}{}_{x,x+n+1} \perp \mathbb{I}^{x,n}_{n+1,n+1} \qquad (3.41b)$$

(as it follows from (3.18a) with n replaced by $n+1$, and $v=n+1$; and from (3.22b) with n replaced by $n+1$).

The four types of the 'local' order-update recursions, indicated in the diagram (3.40), can be interpreted as follows:

LL-recursion: 'local' order-update for the submatrices ${}^{1}A^{x}_{n,o}$ and $B^{x+1,n-1}_{n,o}$

LB-recursions: 'local' order-updates for the submatrices

$${}^{2}A^{x}_{n,m} \text{ and } {}^{1}B^{x+1,n-1}_{n,m} \quad (m=1,\dots,n)$$
$${}^{2}A^{x}_{n,n+1} \text{ and } {}^{1}B^{x+1,n}_{n,n+1} \quad (m=n+1)$$

BL-recursions: 'local' order-updates for the submatrices

$${}^{1}A^{x,o}_{n,1} \text{ and } {}^{2}B^{x}_{n,1} \quad (v=0)$$
$${}^{1}A^{x,v}_{n,v+1} \text{ and } {}^{2}B^{x,v-1}_{n,v+1} \quad (v=1,\dots,n)$$

BB-recursions: 'local' order-updates for the submatrices

$${}^{2}A^{x,o}_{n,m} \text{ and } {}^{2}B^{x}_{n,m} \quad (v=0 \text{ and } m=2,\dots,n+2)$$
$${}^{2}A^{x,v}_{n,m} \text{ and } {}^{2}B^{x,v-1}_{n,m} \quad (v=1,\dots,n \text{ and } m=v+2,\dots,v+n+2)$$

As an example, we shall discuss in some detail the LL 'local' order-update recursion while the remaining LB, BL and BB recursions are presented in Appendix 2.

PROPOSITION 3.1 (The LL normalized recursion)

Given the L-forward (3.16) and L-backward (3.20) approximation errors, we have the following recurrence relations

$$A^{x}_{n,1} = \{1-[\rho^{x}_{n,1}]^2\}^{-\frac{1}{2}}\{[A^{x}_{n,o} \quad 0_{n+1}] - \rho^{x}_{n,1}[0_o \quad B^{x+1,n-1}_{n,o}]\} \quad (3.42a)$$

$$B^{x}_{n,1} = \{1-[\rho^{x}_{n,1}]^2\}^{-\frac{1}{2}}\{-\rho^{x}_{n,1}[A^{x}_{n,o} \quad 0_{n+1}] - [0_o \quad B^{x+1,n-1}_{n,o}]\} \quad (3.42b)$$

where

$$\rho^{x}_{n,1} = ([A^{x}_{n,o} \quad 0_{n+1}],[0_o \quad B^{x+1,n-1}_{n,o}])_{\mathbf{I}^{x}_{n,1}} \quad (3.42c)$$

PROOF.

LL-forward recursion

Let us consider

$$\mathbf{I}^{x}_{n,1} = v\{I^{x}_{n,1}\}$$

This subspace can be rewritten, in accordance with (3.7), as

$$\mathbf{I}^{x}_{n,1} = v\{1_x, I^{x+1,n-1}_{n,o}\} = v\{1_x, I^{x+1,n-1}_{n-1,n}, 1_{x+n+1}\}$$

so that it contains the 'new' uni-variate element 1_{x+n+1}, in comparison with $\mathbf{I}^{x}_{n,o}$ (3.15). The L-forward estimate of the 1_x with updated 'uni-variate' submatrix; i.e., $\hat{1}^{x}_{n,1}$, will be expressed as

$$\hat{1}^{x}_{n,1} = P^{x+1,n-1}_{\mathbf{I};n,o}\, 1_x \in v\{\mathbf{I}^{x+1,n-1}_{n,o}\}$$

The error $A^x_{n,1}$, associated with $\hat{1}^x_{n,1}$, will be expressed as

$$A^x_{n,1} = P^{\perp\ x+1,n-1}_{\mathrm{II};n,o}\ 1_x = 1_x - [\,0_o\ \ \hat{1}^x_{n,1}\,] \perp \mathrm{II}^{x+1,n-1}_{n,o}$$

and $D\,A^x_{n,1} = L^x_{n,1} = L^x_{n,o} \cup \{x+n+1\}$, in accordance with (3.7). From (3.25b) we obtain

$$A^x_{n,1} = (I - P^{x+1,n-1}_{\mathrm{II};n-1,n} - P^{x+1,n-1}_{B;n,o})\ 1_x =$$

$$= (I - P^{x+1,n-1}_{\mathrm{II};n-1,n})\ 1_x - P^{x+1,n-1}_{B;n,o}\ 1_x =$$

$$= 1_x - [0_o\ \ \hat{1}^x_{n,o}\ \ 0_{n+1}] - P^{x+1,n-1}_{B;n,o}\ 1_x$$

Therefore,

$$A^x_{n,1} = [\,A^x_{n,o}\ \ 0_{n+1}] - P^{x+1,n-1}_{B;n,o}\ 1_x$$

with $A^x_{n,o}$ expressed by (3.16a). Since $DB^{x+1,n-1}_{n,o} = L^{x+1,n-1}_{n,o}$ but $D1_x = L^x_{n,1} = \{x\} \cup L^{x+1,n-1}_{n,o}$, we have

$$A^x_{n,1} = [A^x_{n,o}\ \ 0_{n+1}] - (1_x, [\,0_o\ \ B^{x+1,n-1}_{n,o}])_{\mathrm{II}^x_{n,1}}\ [0_o\ \ B^{x+1,n-1}_{n,o}]$$

Since $\hat{1}^x_{n,o} \in \mathrm{II}^{x+1,n-1}_{n-1,n}$ but $\perp B^{x+1,n-1}_{n,o}$ (in accordance with (3.20)), we obtain

$$A^x_{n,1} = [A^x_{n,o}\ \ 0_{n+1}] - ([A^x_{n,o}\ \ 0_{n+1}], [0_o\ \ B^{x+1,n-1}_{n,o}])_{\mathrm{II}^x_{n,1}}\ [0_o\ \ B^{x+1,n-1}_{n,o}]$$

According to (3.16), we get

$$([A^x_{n,o}\ \ 0_{n+1}], [\,0_o\ \ B^{x+1,n-1}_{n,o}])_{\mathrm{II}^x_{n,1}} = \|A^x_{n,o}\|\,\rho^x_{n,1}$$

where we defined

$$\rho^{x}_{n,1} \stackrel{\Delta}{=} ([A^{x}_{n,o} \quad 0_{n+1}] , [0_{o} \quad B^{x+1,n-1}_{n,o}])_{\mathbb{I}^{x}_{n,1}}$$

and we can observe that $|\rho^{x}_{n,1}| \leq 1$. Consequently,

$$A^{x}_{n,1} = \|A^{x}_{n,o}\| \{[A^{x}_{n,o} \quad 0_{n+1}] - \rho^{x}_{n,1} [0_{o} \quad B^{x+1,n-1}_{n,o}]\}$$

Renormalizing $A^{x}_{n,1}$, we get

$$A^{x}_{n,1} = \|A^{x}_{n,1}\|^{-1} \|A^{x}_{n,o}\| \{[A^{x}_{n,o} \quad 0_{n+1}] - \rho^{x}_{n,1} [0_{o} \quad B^{x+1,n-1}_{n,o}]\}$$

Finally, observing that

$$(A^{x}_{n,1}, A^{x}_{n,1})_{\mathbb{I}^{x}_{n,1}} = 1 = \|A^{x}_{n,1}\|^{-2} \|A^{x}_{n,o}\|^{2} (1-[\rho^{x}_{n,1}]^{2})$$

we obtain the normalized LL forward recursion (3.42a).

<u>LL-backward recursion</u>

Let us rewrite $\mathbb{I}^{x}_{n,1}$, in accordance with (3.7), as

$$\mathbb{I}^{x}_{n,1} = \vee \{I^{x}_{n,o}, 1_{x+n+1}\} = \vee \{1_{x}, I^{x+1,n-1}_{n-1,n}, 1_{x+n+1}\}$$

so that it contains the 'new' element 1_{x}, in comparison with $\mathbb{I}^{x+1,n-1}_{n,o}$ (3.19). The L-backward estimate $\check{1}^{x}_{n,1}$ of the 1_{x+n+1} is given by

$$\check{1}^{x}_{n,1} = P^{x}_{\mathbb{I};n,o} 1_{x+n+1} \in \vee\{\mathbb{I}^{x}_{n,o}\}$$

The approximation error, corresponding to that estimate, is then

$$B^{x}_{n,1} = P^{\perp x}_{\mathbb{I};n,o} 1_{x+n+1} = 1_{x+n+1} - [\check{1}^{x}_{n,1} \quad 0_{n+1}] \perp \mathbb{I}^{x}_{n,o}$$

and $D\,\mathcal{B}^{x}_{n,1} = L^{x}_{n,1} = \{x\} \cup L^{x+1,n-1}_{n,o}$, in accordance with (3.7). From (3.23b) we get

$$\mathcal{B}^{x}_{n,1} = (I - P^{x+1,n-1}_{\mathbb{I};n-1,n} - P^{x}_{A;n,o})\, 1_{x+n+1} =$$

$$= (I - P^{x+1,n-1}_{\mathbb{I};n-1,n})\, 1_{x+n+1} - P^{x}_{A;n,o}\, 1_{x+n+1} =$$

$$= 1_{x+n+1} - [0_o \quad \check{1}^{x+1,n-1}_{n,o} \quad 0_{n+1}] - P^{x}_{A;n,o}\, 1_{x+n+1}$$

Consequently,

$$\mathcal{B}^{x}_{n,1} = [\,0_o \quad \mathcal{B}^{x+1,n-1}_{n,o}\,] - P^{x}_{A;n,o}\, 1_{x+n+1}$$

where $\mathcal{B}^{x+1,n-1}_{n,o}$ is given by (3.20). Since $DA^{x}_{n,o} = L^{x}_{n,o}$ but $D1_{x+n+1} = L^{x}_{n,1} = L^{x}_{n,o} \cup \{\, x+n+1\}$, we have

$$\mathcal{B}^{x}_{n,1} = [0_o \;\; \mathcal{B}^{x+1,n-1}_{n,o}] - (1_{x+n+1}, [A^{x}_{n,o} \quad 0_{n+1}])_{\mathbb{I}^{x}_{n,1}} [A^{x}_{n,o} \quad 0_{n+1}]$$

Since $\check{1}^{x+1,n-1}_{n,o} \in \mathbb{I}^{x+1,n-1}_{n-1,n}$ but $\perp A^{x}_{n,o}$ (according to (3.16) and (3.20)), we obtain

$$\mathcal{B}^{x}_{n,1} = [0_o \quad \mathcal{B}^{x+1,n-1}_{n,o}] - ([0_o \;\; \mathcal{B}^{x+1,n-1}_{n,o}], [A^{x}_{n,o} \quad 0_{n+1}])_{\mathbb{I}^{x}_{n,1}} [A^{x}_{n,o} \quad 0_{n+1}] =$$

$$= \|\mathcal{B}^{x+1,n-1}_{n,o}\| \; \{[0_o \quad \mathcal{B}^{x+1,n-1}_{n,o}] - \rho^{x}_{n,1} [A^{x}_{n,o} \quad 0_{n+1}]\}$$

Renormalizing $\mathcal{B}^{x}_{n,1}$, and observing that

$$(\mathcal{B}^{x}_{n,1}, \mathcal{B}^{x}_{n,1})_{\mathbb{I}^{x}_{n,1}} = 1 = \|\mathcal{B}^{x}_{n,1}\|^{-2}\, \|\mathcal{B}^{x+1,n-1}_{n,o}\|^{2}\, (1-[\rho^{x}_{n,1}]^{2})$$

we obtain the normalized LL backward recursion (3.42b).

The remaining 'local' order-update recursions LB, BL and BB , can be obtained in a similar manner, and are summarized in Appendix 2. We can observe that these recursions will constitute the generalized nonlinear Levinson algorithm, introduced algebraically in Zarzycki and Dewilde (1983a); Zarzycki (1983). Here, the generalized nonlinear Levinson algorithm has been introduced geometrically, using projection method, and can therefore be interpreted as a method for (Gram-Schmidt) orthogonalization of the basis of the space of the generalized coefficient-matrices. We notice that, neglecting all 'bi-variate' terms (i.e., the entries of the two-indexed submatrices $^{2}A_{...}$ and $^{2}B_{...}$) and, consequently, reducing the scheme to the LL recursions only, we immediately obtain the generalized (nonstationary) linear Levinson algorithm (associated with second-order sequences), reported in Deprettere and Lie (1980).

The parameters ρ, computed by the algorithm, can be interpreted as the reflection coefficients. For example, the $\rho^{x}_{n,1}$ (3.42c) is numerically equal to the LL reflection coefficient computed by the algorithm presented in Zarzycki and Dewilde (1983a) and/or by the time-variant nonlinear ladder-filter algorithm presented in Zarzycki (1984b), and similarly for the reflection coefficients computed by the LB,BL and BB recursions. This follows from the isomorphism (2.47).

Considering the 'local' order-update recursions, we can observe that during each 'global' order-update step $n \rightarrow n+1$, executed at the x-labeled 'level' (where $x=0,\ldots,N-n$ and $n=0,\ldots,N$), the following set of the reflection coefficients is computed by the nonlinear Levinson algorithm:

$$\begin{array}{|c|c|cccc|}
\hline
 & \rho^{x,n}_{n+1,o} & \rho^{x,n}_{n+1,1} & \cdots & \rho^{x,n}_{n+1,n} & \rho^{x,n}_{n+1,n+1} \\
 & \rho^{x,n-1}_{n,n+1} & \rho^{x,n-1}_{n+1,o} & \cdots & \rho^{x,n-1}_{n+1,n-1} & \rho^{x,n-1}_{n+1,n} \\
\mathrm{BL} & \vdots & \vdots & \mathrm{BB} & \vdots & \vdots \\
 & \rho^{x,o}_{n,2} & \rho^{x,o}_{n,3} & \cdots & \rho^{x,o}_{n+1,o} & \rho^{x,o}_{n+1,1} \\
\hline
 & \rho^{x}_{n,1} & \rho^{x}_{n,2} & \cdots & \rho^{x}_{n,n+1} & \rho^{x}_{n+1,o} \\
\mathrm{LL} & & & \mathrm{LB} & & \\
\hline
\end{array} \qquad (3.43)$$

3.2.7 Optimum approximation of the M-D impulse responses

Considering (3.39b), we obtain the following orthogonal expansion for the N-th 'global' order L-forward approximation error $A^{o}_{N,o}$ (associated with the desired N-th order estimate $\hat{1}^{o}_{N,o}$ of the 1_o)

$$A^{o}_{N,o} = P^{\perp\; 1,N-1}_{\mathrm{II}\,;N-1,N}\; 1_o =$$

$$= 1_o - \rho^{o}_{o,1} B^{1}_{o,o} - \sum_{n=1}^{N-1} \rho^{o}_{n,1} B^{1,n-1}_{n,o} + \qquad (3.44a)$$

$$- \sum_{n=1}^{N-1} \sum_{m=1}^{n} \rho^{o}_{n,m+1} B^{1,n-1}_{n,m} - \sum_{n=0}^{N-1} \rho^{o}_{n+1,o} B^{1,n}_{n,n+1} \qquad (3.44b)$$

Interpreting $A^{o}_{N,o}$ as the set of the M-D impulse responses of the optimum (second-degree) nonlinear approximate prediction filter, we notice that (3.44) is actually the generalized Fourier series, generated by the element 1_o (equivalently - by the random variable y_o, under the isomorphism (2.47)), in the space of the generalized matrices. In other words, this is the desired expansion (3.39c). We also remark that the reflaction coefficients, computed by the nonlinear prediction filter algorithm, are actually the generalized Fourier coefficients in the expansion (3.39c) and/or (3.44).

We can therefore conclude that the nonlinear prediction filter algorithm, introduced in this chapter and equivalent to the generalized nonlinear Levinson algorithm, can be interpreted as the method for recursive (actually Gram-Schmidt) orthogonalization of the basis in the space of the generalized matrices, resulting in the orthonormal expansion for the desired estimate. In other words, the nonlinear least-squares prediction problem for higher-order stochastic sequences, can be equivalently considered as the problem of optimum ON approximation of the set of M-D impulse responses of the desired prediction filter. Moreover, the nonlinear filter algorithm, presented here, can be treated as an effective method for computation of the nonlinear orthogonal approximate filter of the Volterra-Wiener class, since the ON expansion (3.44) is actually an explicite form of the Fourier expansion (1.8) if $M=2$.

Let us observe that (3.44) will reduce to the Fourier expansion for the 1-D impulse response of the optimum linear prediction filter, if all 'bi-variate' terms are neglected. In the next paragraph we will see that estimation accuracy will be improved in the nonlinear case (in comparison with the linear approach) by the existence of the second component (3.44b) in the ON expansion.

3.2.8 Estimation accuracy

Recalling (3.42a), and comparing with equation (i) in the proof of Proposition 3.1, we notice that the norms of the L-forward approximation errors in the LL 'local' order-update recursion can be related as follows

$$\|A^x_{n,1}\| = \|A^x_{n,o}\| \left(1-[\rho^x_{n,1}]^2\right)^{\frac{1}{2}} \tag{3.45a}$$

Considering the LB 'local' recursions of Appendix 2, we can show in a similar way that

$$\|A^x_{n,m}\| = \|A^x_{n,m-1}\| \left(1-[\rho^x_{n,m}]^2\right)^{\frac{1}{2}} \quad , \quad m=2,\ldots,n+1 \tag{3.45b}$$

$$\|A^x_{n,n+2}\| = \|A^x_{n+1,o}\| = \|A^x_{n,n+1}\| \left(1-[\rho^x_{n+1,o}]^2\right)^{\frac{1}{2}} \quad , \quad m=n+2 \tag{3.45c}$$

Combining (3.45a)-(3.45c), we obtain the error-norm relation in the 'global' order-update step $n \to n+1$ as

$$\|A^x_{n+1,o}\| = \|A^x_{n,o}\| \, R^x_{LL;n+1} \, R^x_{LB;n+1} \tag{3.46a}$$

where $R^x_{LL;n+1}$ is the factor associated with the LL 'local' recursion

$$R^x_{LL;n+1} = \left(1-[\rho^x_{n,1}]^2\right)^{\frac{1}{2}} \tag{3.46b}$$

while $R^x_{LB;n+1}$ is the factor corresponding to the LB 'local' recursions (3.45b,c), and is expressed as

$$R^x_{LB;n+1} = \left[\prod_{m=2}^{n+1} \left(1-[\rho^x_{n,m}]^2\right)^{\frac{1}{2}}\right] \left(1-[\rho^x_{n+1,o}]^2\right)^{\frac{1}{2}} \tag{3.46c}$$

Thus, equations (3.46) express the error-norm reduction in the 'global' order-update step, in the (second-degree) nonlinear Levinson scheme. In the linear Levinson case, (3.46) would be replaced by

$$\|A^x_{n+1}\| = \|A^x_n\| \left(1-[\rho^x_{n+1}]^2\right)^{\frac{1}{2}} \tag{3.47}$$

see also Deprettere and Lie (1980). Therefore, better estimation accuracy is achieved due to the factor (3.46c), associated with the LB 'local' recursions.

Considering (3.46) for $x=0,\ldots,N-x$, we obtain

$$\| A^{x}_{N-x,o} \| = \| A^{x}_{o,o} \| R^{x}_{LL} R^{x}_{LB} \tag{3.48a}$$

where R^{x}_{LL} is the factor associated with all LL recursions executed at the x-labeled 'level' in the N-th order scheme, and is given by

$$R^{x}_{LL} = \prod_{n=0}^{N-x-1} R^{x}_{LL;n+1} = \prod_{n=0}^{N-x-1} (1-[\rho^{x}_{n,1}]^2)^{\frac{1}{2}} . \tag{3.48b}$$

while the R^{x}_{LB} factor is associated with all LB 'local' steps, executed at that 'level', and can be written as

$$R^{x}_{LB} = \prod_{n=0}^{N-x-1} R^{x}_{LB;n+1} =$$

$$= \prod_{n=0}^{N-x-1} \left[\prod_{m=2}^{n+1} (1-[\rho^{x}_{n,m}]^2)^{\frac{1}{2}} \right] (1-[\rho^{x}_{n+1,o}]^2)^{\frac{1}{2}} \tag{3.48c}$$

Equations (3.48) relate the error-norms in the N-th order, second-degree nonlinear scheme, at the x-labeled 'level'. Interpreting the parameter x as a backward shift (i.e., time-delay) from the reference point $t=0$, the relations (3.48) will correspond to the time-instant $t=x$. Since the desired estimate 1_o corresponds to $t=0$, estimation accuracy for that estimate will be expressed by (3.48) with $x=0$.

In the linear N-th order scheme, we get (according to (3.47) considered for $n=0,\ldots,N-x$)

$$\| A^{x}_{N-x} \| = \| A^{x}_{o} \| \prod_{n=0}^{N-x-1} (1-[\rho^{x}_{n+1}]^2)^{\frac{1}{2}} \tag{3.49}$$

so that better estimation accuracy in the nonlinear scheme will be achieved due to the factor (3.48c) associated with the LB (i.e., nonlinear) recursions (since the norms of all reflection coefficients computed by the nonlinear filter algorithm are less than 1).

The generalized nonlinear Levinson algorithm, introduced here geometrically using projection method in the space of the generalized matrices, and interpreted in terms of the optimum orthonormal approximation of the set of M-D impulse responses, will imply the time-variant orthogonal nonlinear digital prediction filter. Its structure and properties will be discussed in paragraph 3.4.

3.3 Nonlinear filter algorithm: transform-domain approach

Let for $n=0,\dots,N$ and $x=0,\dots,N-n$

$$Z^{x}_{n,m} = [{}^{1}Z^{x}_{n,m} \quad {}^{2}Z^{x}_{n,m}] = [z_{j_1} \quad z_{j_1 j_2}] \quad (j_1,j_2)\in L^{x}_{n,m} \tag{3.50a}$$

where

$$z_{j_1} = [z_1^{j_1-x} \quad 0] \quad ; \quad z_{j_1 j_2} = [0 \quad z_1^{j_1-x}\, z_2^{j_2-x}] \tag{3.50b}$$

and similarly for $Z^{x,v}_{n,m}$. Then we can consider the following subspaces

$$\mathbf{Z}^{x}_{n,m} = \vee\{Z^{x}_{n,m}\} \quad ; \quad \mathbf{Z}^{x,v}_{n,m} = \vee\{Z^{x,v}_{n,m}\} \tag{3.51}$$

We shall denote by $P^{x}_{\mathbf{Z};n,m}$ $(P^{x,v}_{\mathbf{Z};n,m})$ the orthogonal projection operator, taking projection on the subspace $\mathbf{Z}^{x}_{n,m}$ $(\mathbf{Z}^{x,v}_{n,m})$. We notice that $\mathbf{Z}^{1,N-1}_{N-1,N}$ will be the 'estimation subspace' (see (2.44) with $M=2$), under the isomorphism (2.47), while the 'biggest' subspace will actually be $\mathbf{Z}^{o,N}_{N,N+1}$.

Each element $\Phi^{x}_{n,m}$ from the subspace $\mathbf{Z}^{x}_{n,m}$ will be expressed as

$$\Phi^{x}_{n,m}(Z) = [\Phi^{x}_{n,m}(z_1) \quad \Phi^{x}_{n,m}(z_1,z_2)] \tag{3.52a}$$

where

$$\Phi^{x}_{n,m}(z_1) = \sum_{j_1} f^{x}_{n,m;j_1-x}\, z_1^{j_1-x} \tag{3.52b}$$

and

$$\Phi^{x}_{n,m}(z_1,z_2) = \sum_{j_1}\sum_{j_2} f^{x}_{n,m;j_1-x,j_2-x}\, z_1^{j_1-x} z_2^{j_2-x} \tag{3.52c}$$

where the summations in (3.52b,c) are over the uni-variate, respectively, over the bi-variate parts of the index-set $L^{x}_{n,m}$. Following (2.46a), we introduce inner-product on $\mathbf{Z}^{x}_{n,m}$ as

$$(\Phi^{x}_{n,m}(Z), \Psi^{x}_{n,m}(Z))_{\mathbf{Z}} =$$

$$= \frac{1}{(2\pi)^2}\int d\theta_1 \int d\omega_1\, \Phi^{x}_{n,m}(z_1)\, {}^{1\oplus 1}W(e^{i\theta_1},e^{i\omega_1})\, \bar{\Psi}^{x}_{n,m}(w_1) \; +$$

$$+ \frac{1}{(2\pi)^3}\int d\theta_1 \int d\theta_2 \int d\omega_1\, \Phi^{x}_{n,m}(z_1,z_2)\, {}^{2\oplus 1}W(e^{i\theta_1},e^{i\theta_2},e^{i\omega_1})\, \bar{\Psi}^{x}_{n,m}(w_1) \; +$$

$$+ \frac{1}{(2\pi)^3}\int d\theta_1 \int d\omega_1 \int d\omega_2\, \Phi^{x}_{n,m}(z_1)\, {}^{1\oplus 2}W(e^{i\theta_1},e^{i\omega_1},e^{i\omega_2})\, \bar{\Psi}^{x}_{n,m}(w_1,w_2) \; +$$

$$+ \frac{1}{(2\pi)^4}\int d\theta_1 \int d\theta_2 \int d\omega_1 \int d\omega_2\, \Phi^{x}_{n,m}(z_1,z_2)\, {}^{2\oplus 2}W(e^{i\theta_1},e^{i\theta_2},e^{i\omega_1},e^{i\omega_2}) \times$$

$$\times \bar{\Psi}^{x}_{n,m}(w_1,w_2) \tag{3.53}$$

where ${}^{m\oplus u}W(e^{i\theta^m},e^{i\omega^u})$, $m,u=1,2$ are the entries of the (2×2) spectral matrix (2.14) of the fourth-order stochastic sequence $\{y\}$. Similar relations can be written for the subspaces $\mathbf{Z}^{x,v}_{n,m}$.

Following the ordering scheme of paragraph 3.1, let us introduce the 'local' subspaces:

L-forward

$$\mathbf{Z}^{x}_{n,o} = \vee \{ z_x , Z^{x+1,n-1}_{n-1,n} \} \tag{3.54a}$$

B-forward (for v=0,...,n)

$$\mathbf{Z}^{x,v}_{n,v+1} = \begin{cases} \vee \{ z_{x,x} , Z^{x}_{n,o} \} & , \quad v=0 \\ \vee \{ z_{x,x+v} , Z^{x,v-1}_{n,v} \} & , \quad v=1,\dots,n \end{cases} \tag{3.54b}$$

L-backward

$$\mathbf{Z}^{x,n-1}_{n,o} = \vee \{ Z^{x,n-1}_{n-1,n} , z_{x+n+1} \} \tag{3.55a}$$

B-backward (for m=1,...,n+1)

$$\mathbf{Z}^{x,n-1}_{n,m} = \vee \{ Z^{x,n-1}_{n,m-1} , z_{x+n+1-m,x+n} \} \quad , \quad m=1,\dots,n \tag{3.55b}$$

$$\mathbf{Z}^{x,n}_{n,n+1} = \vee \{ Z^{x,n-1}_{n,n} , z_{x,x+n} \} \quad , \quad m=n+1 \tag{3.55c}$$

3.3.1 'Local' estimates and errors

Following (3.54) and (3.55), we can introduce the L- and B-, forward and backward estimates and errors of the n-th 'global' order:

L-forward

The L-forward estimate $\hat{z}^{x}_{n,o}$ of the element z_x will be defined as

$$\hat{z}^{x}_{n,o} \stackrel{\Delta}{=} P_{\mathbf{Z};n-1,n}^{x+1,n-1} z_x \in \mathbf{Z}^{x+1,n-1}_{n-1,n} \tag{3.56a}$$

while the corresponding L-forward approximation error will be

$$\Delta^{x}_{n,o} \overset{\Delta}{=} P^{\perp\, x+1,n-1}_{\mathbf{Z};n-1,n}\, z_{x} \perp \mathbf{Z}^{x+1,n-1}_{n-1,n} \tag{3.56b}$$

Its normalized version will be expressed as

$$A^{x}_{n,o}(Z) = \Delta^{x}_{n,o}\, \|\Delta^{x}_{n,o}\|^{-1}_{\mathbf{Z}} = [A^{x}_{n,o}(z_1) \quad A^{x}_{n,o}(z_1,z_2)] \tag{3.56c}$$

We can observe that the estimate $\hat{z}^{x}_{n,o}$ and the errors $\Delta^{x}_{n,o}$, $A^{x}_{n,o}(Z)$ are precisely the ${}^{M}\hat{z}_{N;o}$, ${}^{M}\Delta_{N;o}$ and ${}^{M}\mathbf{A}_{N;o}$ (2.57), if $M=2$, $x=0$ and $n=N$.

<u>B-forward</u> (for $v=0,\ldots,n$)

The B-forward estimates $\hat{z}^{x,v}_{n,v+1}$ of the elements $z_{x,x+v}$ will be

$$\hat{z}^{x,v}_{n,v+1} \overset{\Delta}{=} \begin{cases} P^{x}_{\mathbf{Z};n,o}\, z_{x,x} \in \mathbf{Z}^{x}_{n,o} , & v=0 \\ P^{x,v-1}_{\mathbf{Z};n,v}\, z_{x,x+v} \in \mathbf{Z}^{x,v-1}_{n,v} , & v=1,\ldots,n \end{cases} \tag{3.57a}$$

while the approximation errors are given by

$$\Delta^{x,v}_{n,v+1} \overset{\Delta}{=} \begin{cases} P^{\perp\, x}_{\mathbf{Z};n,o}\, z_{x,x} \perp \mathbf{Z}^{x}_{n,o} , & v=0 \\ P^{\perp\, x,v-1}_{\mathbf{Z};n,v}\, z_{x,x+v} \perp \mathbf{Z}^{x,v-1}_{n,v} , & v=1,\ldots,n \end{cases} \tag{3.57b}$$

or, in a renormalized form,

$$A^{x,v}_{n,v+1}(Z) = \Delta^{x,v}_{n,v+1}\|\Delta^{x,v}_{n,v+1}\|^{-1}_{\mathbf{Z}} = [A^{x,v}_{n,v+1}(z_1) \quad A^{x,v}_{n,v+1}(z_1,z_2)] \tag{3.57c}$$

<u>L-backward</u>

The L-backward estimate of z_{x+n} will be expressed as

$$\check{z}^{x,n-1}_{n,o} \overset{\Delta}{=} P^{\;\;x,n-1}_{\mathbf{Z};n-1,n}\, z_{x+n} \quad \in \quad \mathbf{Z}^{x,n-1}_{n-1,n} \tag{3.58a}$$

and the corresponding L-backward error is given by

$$\Gamma^{x,n-1}_{n,o} \overset{\Delta}{=} P^{\perp\; x,n-1}_{\mathbf{Z};n-1,n}\, z_{x+n} \quad \perp \quad \mathbf{Z}^{x,n-1}_{n-1,n} \tag{3.58b}$$

together with its normalized version

$$B^{x,n-1}_{n,o}(Z) = \Gamma^{x,n-1}_{n,o} \| \Gamma^{x,n-1}_{n,o} \|^{-1}_{\mathbf{Z}} \quad = [B^{x,n-1}_{n,o}(z_1) \quad B^{x,n-1}_{n,o}(z_1,z_2)] \tag{3.58c}$$

<u>B-backward</u> (for $m=1,\dots,n+1$)

The B-backward estimates of the elements $z_{x+n,x+n}, \dots, z_{x,x+n}$ are given by

$$\check{z}^{x,n-1}_{n,m} \overset{\Delta}{=} P^{\;\;x,n-1}_{\mathbf{Z};n,m-1}\, z_{x+n+1-m,x+n} \;\in\; \mathbf{Z}^{x,n-1}_{n,m-1} \quad , \quad m=1,\dots,n \tag{3.59a}$$

$$\check{z}^{x,n}_{n,n+1} \overset{\Delta}{=} P^{\;\;x,n-1}_{\mathbf{Z};n,n}\, z_{x,x+n} \quad \in \quad \mathbf{Z}^{x,n-1}_{n,n} \quad , \quad m=n+1 \tag{3.59b}$$

The corresponding B-backward errors are

$$\Gamma^{x,n-1}_{n,m} \overset{\Delta}{=} P^{\perp\; x,n-1}_{\mathbf{Z};n,m-1}\, z_{x+n+1-m,x+n} \;\perp\; \mathbf{Z}^{x,n-1}_{n,m-1} \quad , \quad m=1,\dots,n \tag{3.59c}$$

$$\Gamma^{x,n}_{n,n+1} \overset{\Delta}{=} P^{\perp\; x,n-1}_{\mathbf{Z};n,n}\, z_{x,x+n} \quad \perp \quad \mathbf{Z}^{x,n-1}_{n,n} \quad , \quad m=n+1 \tag{3.59d}$$

Their normalized versions are $B^{x,n-1}_{n,m}(Z)$ and $B^{x,n}_{n,n+1}(Z)$, respectively, and are expressed similarly as in (3.58).

Let us observe that the L- and B-, forward and backward estimates $\hat{z}^{x}_{n,m}$, $\hat{z}^{x,v}_{n,m}$, $\check{z}^{x,v}_{n,m}$ and errors $A^{x}_{n,m}(Z)$, $A^{x,v}_{n,m}(Z)$, $B^{x,v}_{n,m}(Z)$ are the representatives of the estimates $\hat{1}^{x}_{n,m}$, $\hat{1}^{x,v}_{n,m}$, $\check{1}^{x,v}_{n,m}$ and of the errors $A^{x}_{n,m}$, $A^{x,v}_{n,m}$, $B^{x,v}_{n,m}$ of paragraph 3.2, under the isomorphism (2.47).

3.3.2 Decomposition of subspaces, ON bases and M-D Fourier expansion

Let $\Pi^{x}_{A;n,m}$, $\Pi^{x,v}_{A;n,m}$, $\Pi^{x,v}_{B;n,m}$ denote the orthogonal projection operators on the subspaces spanned by $A^{x}_{n,m}(Z)$, $A^{x,v}_{n,m}(Z)$, $B^{x,v}_{n,m}(Z)$, respectively. The orthogonality conditions (3.56b),(3.57b),(3.58b) and (3.59c,d) will imply the following orthogonal decompositions of subspaces and of projection operators:

L-forward

$$\mathbf{Z}^{x}_{n,o} = \mathbf{Z}^{x+1,n-1}_{n-1,n} \oplus \vee\{A^{x}_{n,o}(Z)\} \tag{3.60a}$$

$$P^{x}_{\mathbf{Z};n,o} = P^{x+1,n-1}_{\mathbf{Z};n-1,n} + \Pi^{x}_{A;n,o} \tag{3.60b}$$

B-forward (for v=0,...,n)

$$\mathbf{Z}^{x,v}_{n,v+1} = \begin{cases} \mathbf{Z}^{x}_{n,o} \oplus \vee\{A^{x,o}_{n,1}(Z)\} & , \quad v=0 \\ \mathbf{Z}^{x,v-1}_{n,v} \oplus \vee\{A^{x,v}_{n,v+1}(Z)\} & , \quad v=1,\ldots,n \end{cases} \tag{3.61a}$$

$$P^{x,v}_{\mathbf{Z};n,v+1} = \begin{cases} P^{x}_{\mathbf{Z};n,o} + \Pi^{x,o}_{A;n,1} & , \quad v=0 \\ P^{x,v-1}_{\mathbf{Z};n,v} + \Pi^{x,v}_{A;n,v+1} & , \quad v=1,\ldots,n \end{cases} \tag{3.61b}$$

L-backward

$$\mathbf{Z}^{x,n-1}_{n,o} = \mathbf{Z}^{x,n-1}_{n-1,n} \oplus \vee \{ B^{x,n-1}_{n,o}(Z) \} \qquad (3.62a)$$

$$P^{x,n-1}_{\mathbf{Z};n,o} = P^{x,n-1}_{\mathbf{Z};n-1,n} + \Pi^{x,n-1}_{B;n,o} \qquad (3.62b)$$

B-backward (for $m=1,\ldots,n+1$)

$$\mathbf{Z}^{x,n-1}_{n,m} = \mathbf{Z}^{x,n-1}_{n,m-1} \oplus \vee \{B^{x,n-1}_{n,m}(Z)\} \quad , \quad m=1,\ldots,n \qquad (3.63a)$$

$$\mathbf{Z}^{x,n}_{n,n+1} = \mathbf{Z}^{x,n-1}_{n,n} \oplus \vee \{B^{x,n}_{n,n+1}(Z)\} \quad , \quad m=n+1 \qquad (3.63b)$$

$$P^{x,n-1}_{\mathbf{Z};n,m} = P^{x,n-1}_{\mathbf{Z};n,m-1} + \Pi^{x,n-1}_{B;n,m} \quad , \quad m=1,\ldots,n \qquad (3.63c)$$

$$P^{x,n}_{\mathbf{Z};n,n+1} = P^{x,n-1}_{\mathbf{Z};n,n} + \Pi^{x,n}_{B;n,n+1} \quad , \quad m=n+1 \qquad (3.63d)$$

We can observe that the entries of

$$\mathbf{A}^{x}_{n}(Z) = [\, A^{x}_{n,o}(Z) \quad A^{x,o}_{n,1}(Z) \quad \cdots \quad A^{x,n}_{n,n+1}(Z) \,] \qquad (3.64a)$$

as well as of

$$\mathbf{B}^{x}_{n}(Z) = [\, B^{x,n-1}_{n,o}(Z) \quad \cdots \quad B^{x,n-1}_{n,n}(Z) \quad B^{x,n}_{n,n+1}(Z) \,] \qquad (3.64b)$$

will form the ON sets in the space of the generalized z-polynomials, much like their time-domain counterparts form the ON sets in the space of the generalized matrices. The sets (3.64) are actually the orthonormalized versions of

$$[\, z_{x} \quad z_{x,x} \quad \cdots \quad z_{x,x+n} \,] \qquad (3.64c)$$

and of

$$[z_{x+n} \quad z_{x+n,x+n} \quad \cdots \quad z_{x,x+n}] \tag{3.64d}$$

respectively. Therefore, we can write the following 'global' forward and backward orthogonal decompositions of subspaces

$$\mathbf{Z}^{x,n}_{n,n+1} = \mathbf{Z}^{x+1,n-1}_{n-1,n} \oplus \vee\{\mathbf{A}^{x}_{n}(Z)\} \tag{3.65a}$$

$$\mathbf{Z}^{x,n}_{n,n+1} = \mathbf{Z}^{x,n-1}_{n-1,n} \oplus \vee\{\mathbf{B}^{x}_{n}(Z)\} \tag{3.65b}$$

where

$$\vee\{\mathbf{A}^{x}_{n}(Z)\} = \vee\{A^{x}_{n,o}(Z)\} \oplus \sum_{v=0}^{n} \oplus \vee\{A^{x,v}_{n,v+1}(Z)\} \tag{3.65c}$$

$$\vee\{\mathbf{B}^{x}_{n}(Z)\} = \sum_{m=0}^{n} \oplus \vee\{B^{x,n-1}_{n,m}(Z)\} \oplus \vee\{B^{x,n}_{n,n+1}(Z)\} \tag{3.65d}$$

These orthogonal decomposition of subspaces imply the following 'global' decompositions of projection operators

$$P^{x,n}_{\mathbf{Z};n,n+1} = P^{x+1,n-1}_{\mathbf{Z};n-1,n} + \Pi^{x}_{\mathbf{A};n} \tag{3.65e}$$

$$P^{x,n}_{\mathbf{Z};n,n+1} = P^{x,n-1}_{\mathbf{Z};n-1,n} + \Pi^{x}_{\mathbf{B};n} \tag{3.65f}$$

where

$$\Pi^{x}_{\mathbf{A};n} = \Pi^{x}_{A;n,o} + \sum_{v=0}^{n} \Pi^{x,v}_{A;n,v+1} \tag{3.65g}$$

$$\Pi^{x}_{\mathbf{B};n} = \sum_{m=0}^{n} \Pi^{x,n-1}_{B;n,m} + \Pi^{x,n}_{B;n,n+1} \tag{3.65h}$$

Consequently, we get the following orthogonal decomposition

$$\mathbf{Z}^{o,N}_{N,N+1} = \sum_{x=0}^{N} \oplus \vee\{\mathbf{A}^{x}_{N-x}(Z)\} = \sum_{n=0}^{N} \oplus \vee\{\mathbf{B}^{o}_{N-n}(Z)\} \tag{3.66}$$

Hence, the sets $\mathbb{A}_n^x(Z)$ and $\mathbb{B}_n^x(Z)$ can be interpreted as the 'global' forward, resp. backward ON basis of the space of the generalized, multi-variate z-polynomials, square-integrable on the generalized unit-torus with respect to the spectral matrix of the underlying higher-order (i.e., fourth-order) stochastic sequence $\{y\}$.

We also obtain the following orthogonal decomposition of the 'estimation space'

$$\mathbf{Z}_{N-1,N}^{1,N-1} = \sum_{n=0}^{N-1} \oplus \vee \{\mathbb{B}_n^1(Z)\} \tag{3.67a}$$

which implies

$$P_{\mathbf{Z};N-1,N}^{1,N-1} = \sum_{n=0}^{N-1} \Pi_{\mathbb{B};n}^{1} \tag{3.67b}$$

Let us observe that here the element $z_o = [1 \quad 0]$ will actually be the representative of the y_o and/or of the 1_o in the isomorphism (2.47). Thus, the Fourier series for the z_o will be the transform-domain counterpart of the optimum solution to the nonlinear least-squares prediction problem. Following (3.67b), that ON expansion will be expressed as

$$z_o \sim \sum_{n=0}^{N-1} (z_o, \mathbb{B}_n^1(Z))_{\mathbf{Z}} \mathbb{B}_n^1(Z) \tag{3.67c}$$

We will see later that the above ON series can be interpreted in terms of optimum approximation of the set of transfer functions of the nonlinear prediction filter.

We notice that the ON backward polynomial basis of the space (3.67a) should be derived in order to get the Fourier expansion (3.67c). In the next paragraph we will show that the desired basis is computed by the transform-domain version of the nonlinear prediction filter algorithm.

3.3.3 Order-update recursions

In this paragraph we wish to show that the ON basis of the space of the generalized polynomials can be recursively computed by the transform-domain counterpart of the nonlinear filter algorithm, derived in paragraph 3.2.6. In order to do that, it is sufficient to show that the higher-order sets of the generalized polynomials $\mathbb{A}^{x}_{n+1}(Z)$ and $\mathbb{B}^{x}_{n+1}(Z)$ can be derived recursively from the lower-order ON sets $\mathbb{A}^{x}_{n}(Z)$ and $\mathbb{B}^{x+1}_{n}(Z)$, preserving the mutual orthonormality of their entries. This can be done by introducing a set of appropriately derived 'local' order-update recursions, executed on the multi-variate z-polynomial entries of the sets $\mathbb{A}^{x}_{n}(Z)$ and $\mathbb{B}^{x+1}_{n}(Z)$, much like in the scheme (3.40). Here, the LL, LB, BL and BB 'local' recursions of that scheme, will be interpreted as follows:

<u>LL-recursion:</u> order-update step for the polynomials

$$A^{x}_{n,o}(z_1) \quad \text{and} \quad B^{x+1,n-1}_{n,o}(z_1)$$

<u>LB-recursions:</u> order-update steps for the polynomials

$$A^{x}_{n,m}(z_1,z_2) \quad \text{and} \quad B^{x+1,n-1}_{n,m}(z_1) \quad , \quad m=1,\dots,n$$
$$A^{x}_{n,n+1}(z_1,z_2) \quad \text{and} \quad B^{x+1,n}_{n,n+1}(z_1) \quad , \quad m=n+1$$

<u>BL-recursions:</u> order-update steps for the polynomials

$$A^{x,o}_{n,1}(z_1) \quad \text{and} \quad B^{x}_{n,1}(z_1,z_2) \quad , \quad v=0$$
$$A^{x,v}_{n,v+1}(z_1) \quad \text{and} \quad B^{x,v-1}_{n,v+1}(z_1,z_2) \quad , \quad v=1,\dots,n$$

<u>BB-recursions:</u> order-update steps for the polynomials

$$A^{x,o}_{n,m}(z_1,z_2) \quad \text{and} \quad B^{x}_{n,m}(z_1,z_2) \quad , \quad v=0 \quad \text{and} \quad m=2,\dots,n+2$$
$$A^{x,v}_{n,m}(z_1,z_2) \quad \text{and} \quad B^{x,v-1}_{n,m}(z_1,z_2) \quad , \quad v=1,\dots,n \quad \text{and}$$

$m=v+1,\dots,v+n+2.$

These 'local' recursions can be introduced by considering the orthogonal decompositions of the subspaces and of projection operators, much like in the time-domain solution considered in the previous paragraph. For example, the LL 'local' recursion is introduced via

PROPOSITION 3.2

Given the L-forward error $A^{x}_{n,o}(Z)$ (3.56c), and the L-backward error $B^{x+1,n-1}_{n,o}(Z)$ (3.58c), we have the following recurrence relations

$$\begin{bmatrix} A^{x}_{n,1}(Z) \\ B^{x}_{n,1}(Z) \end{bmatrix} = \Theta^{x}_{n,1} \begin{bmatrix} A^{x}_{n,o}(Z) \\ Z\cdot B^{x+1,n-1}_{n,o}(Z) \end{bmatrix} \tag{3.68a}$$

where

$$\Theta^{x}_{n,1} \triangleq (1-[\rho^{x}_{n,1}]^2)^{-\frac{1}{2}} \begin{bmatrix} 1 & -\rho^{x}_{n,1} \\ -\rho^{x}_{n,1} & 1 \end{bmatrix} \tag{3.68b}$$

is the J-unitary factor; i.e. $\Theta^{x}_{n,1} J \tilde{\Theta}^{x}_{n,1} = J$; $J = \begin{bmatrix} 1 & 0 \\ 0 & -1 \end{bmatrix}$, and

$$\rho^{x}_{n,1} = (A^{x}_{n,o}(Z)\, ,\, Z\cdot B^{x+1,n-1}_{n,o}(Z))_{\mathbf{Z}} \tag{3.68c}$$

(in accordance with (3.53)), with

$$Z\cdot B^{x,v}_{n,m}(Z) \triangleq [\, z_1\cdot B^{x,v}_{n,m}(z_1) \qquad z_1 z_2\cdot B^{x,v}_{n,m}(z_1,z_2)\,] \tag{3.68d}$$

PROOF.

Parallels the proof of Proposition 3.1, and is omitted.

We can observe that the reflection coefficient (3.68c) in numerically equal to (3.42c), under the isomorphism (2.47), and that (3.68) is the transform-domain counterpart of the LL Levinson recursion (3.42). The remaining transform-domain versions of the LB, BL and BB recursions can be obtained in a similar way, and are presented in Appendix 2. Considered accordingly together, they will constitute the transform-domain version of the generalized nonlinear Levinson algorithm, which can be interpreted in the transform-domain context, as the method for recursive orthogonalization of the basis in the space of the generalized z-polynomials.

3.3.4 Optimum ON approximation of the set of M-D transfer functions

Recalling (3.67b), and considering the 'local' transform-domain recursions of Appendix 2, we can write the following ON expansion for the error $\Delta^{o}_{N,o}(Z)$

$$\Delta^{o}_{N,o}(Z) = P^{\perp\, 1,N-1}_{\mathbf{Z};N-1,N}\; z_o =$$

$$= z_o - \rho^{o}_{o,1}\, B^{1}_{o,o}(Z) - \sum_{n=1}^{N-1} \rho^{o}_{n,1}\, B^{1,n-1}_{n,o}(Z) + \tag{3.69a}$$

$$+ \sum_{n=1}^{N-1} \sum_{m=1}^{n} \rho^{o}_{n,m+1}\, B^{1,n-1}_{n,m}(Z) + \sum_{n=0}^{N-1} \rho^{o}_{n+1,o}\, B^{1,n}_{n,n+1}(Z) \tag{3.69b}$$

which is the generalized (M-D) Fourier series, generated by the element $z_o = [1 \quad 0\,]$ with respect to the ON backward polynomial basis of the 'estimation space' in the transform-domain. In other words, this is the explicite version of the expansion (3.67c). We notice that (3.69) is equivalent to the Fourier expansion (3.44), and/or to the stochastic functional Fourier series for the nonlinear estimate in the space of the Volterra functional polynomials (see e.g., Zarzycki (1984b)), under the isomorphism

(2.47). We can therefore conclude that the transform-domain version of nonlinear least-squares prediction problem for higher-order stochastic sequences, is actually the optimum ON approximation of the set of M-D transfer functions of the nonlinear prediction filter.

Neglecting all bi-variate terms, this result will reduce to the connection between the time- and transform-domain solutions to the linear least-squares prediction problem, considered in Dewilde, Vieira and Kailath (1978), Dewilde and Dym (1981), Dewilde (1982).

3.4 Nonlinear time-variant ladder-filter

We can observe that the LL 'local' order-update recursion (3.68) can be interpreted as the following LL section

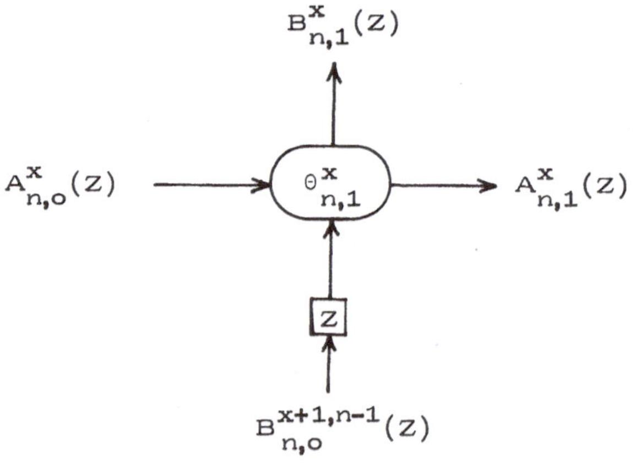

(3.70)

of the nonlinear digital prediction filter. The remaining 'local' order-update recursions can be interpreted in a similar way, yielding the LB, BL and BB sections of that filter (see also Appendix 2). Connected accordingly together, they will constitute the 'global' section $n \rightarrow n+1$ of the nonlinear filter (at the x-labeled 'level'). This 'local' structure of the 'global' filter section is presented in Fig. 3.4 while the 'local' structure of the third-order (N=3) nonlinear prediction ladder-filter is shown in Fig. 3.5.

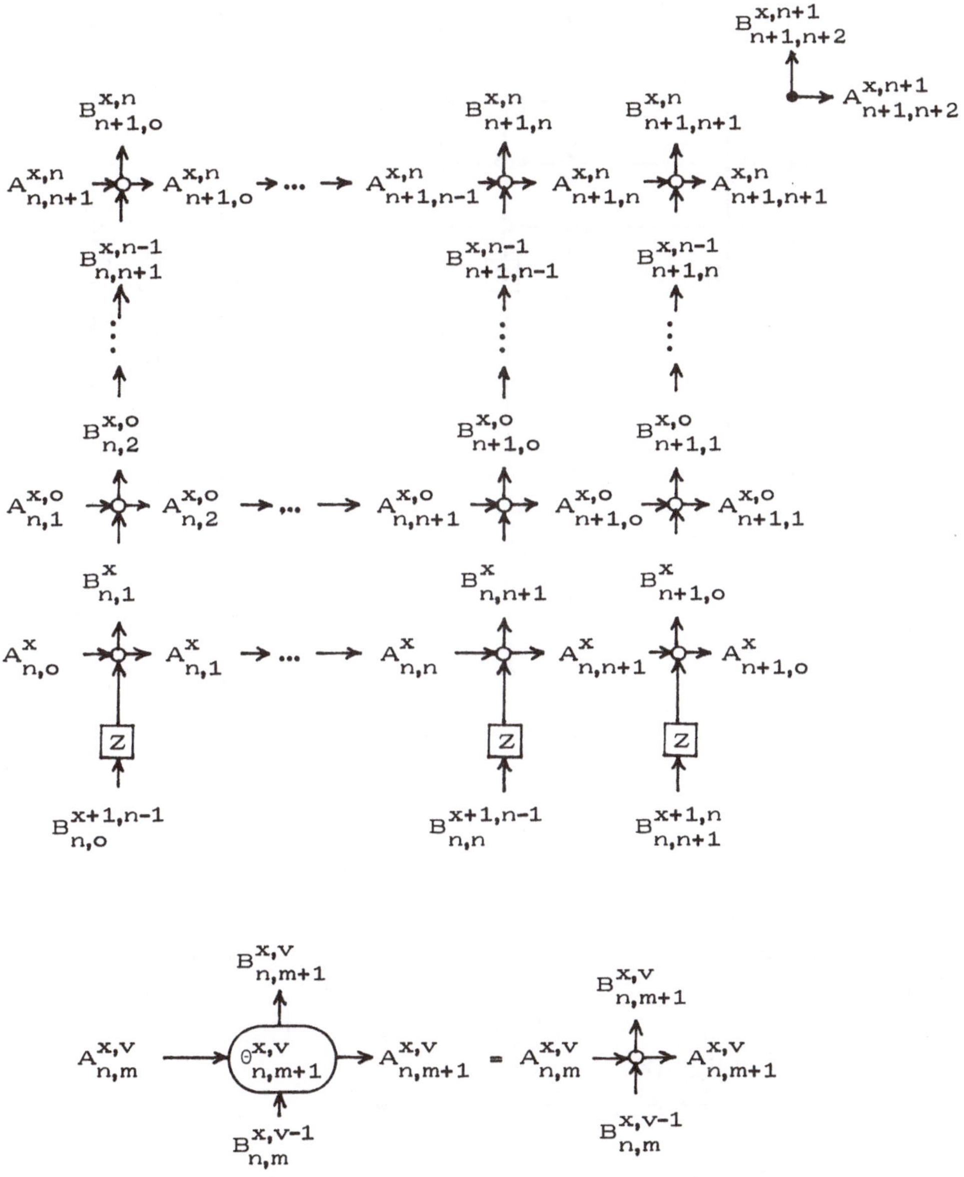

Fig. 3.4 'Local' structure of the 'global' section of the time-variant nonlinear ladder-filter.

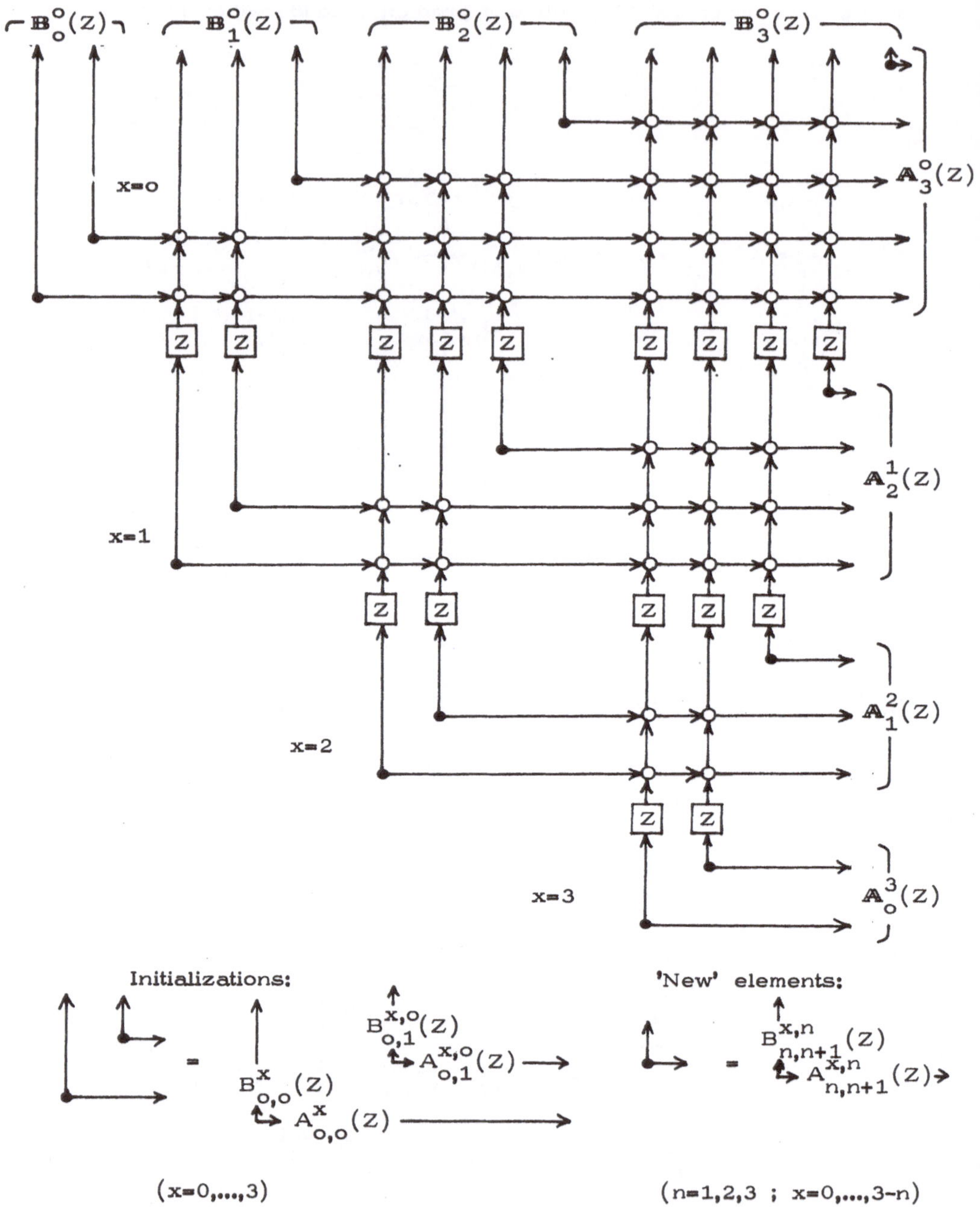

Fig. 3.5 'Local' structure of the third-order (N=3) nonlinear (Levinson) time-variant prediction filter. The symbols O indicate corresponding 'local' Θ-recursions of Fig. 3.4.

Let us observe that the filter structure consists of a cluster of nested 'local' Θ-transformations (actually - Given's rotations) so that the subsequent sets of the forward as well as backward errors will remain mutually orthonormal after each 'local' and, hence, after each 'global' order-update step. (see the subsequent vertical, resp. horizontal solutions in Figs. 3.4 and 3.5). This means that the structure of this nonlinear ladder-filter will satisfy the desired orthogonality requirements, as the filter consists of the 'local' orthogonal and normalized 'local' sections. Each section is associated with the J-lossless Θ-matrix (see Dewilde, Vieira and Kailath (1978), Dewilde and Dym (1981,1984), Dewilde (1982)). These matrices are specified by the reflection coefficients (with norms less than one), being actually the Fourier coefficients.

Since the reflection coefficients (i.e.,the filter 'gains') depend on the parameter x, being the backward shift from the reference point of observation, which can be interpreted as the 'current' time, this nonlinear prediction filter is time-variant. This is implied by the nonstationarity of the higher-order input sequence $\{y\}$, and is reflected in the Hermitian property (not Toeplitz) of the generalized covariance matrix.

We notice that the nonlinear ladder-filter computes the forward as well as backward ON bases, and the solution to the N-th order nonlinear prediction problem (actually considered in the innovations context) is obtained at the 0-labeled 'level' in the filter structure. We also remark that neglecting all nonlinear terms, the filter of Fig. 3.5 will immediately reduce to the generalized (time-variant) linear Levinson filter considered in Deprettere and Lie (1980).

We can conclude that the nonlinear least-squares prediction problem can be solved geometrically, using projection method, in the space of the generalized (block, multi-indexed) coefficient-matrices, and/or in the space of the generalized (block, multi-variate) z-polynomials (provided the higher-order covariance data or, equivalently, the higher-order spectral functions of the underlying stochastic sequence are given). The former approach results in the optimum ON approximation of the set of the M-D impulse responses of the nonlinear prediction filter of the Volterra-Wiener class. In the latter case, the optimum ON polynomial approximation of the set of M-D transfer functions is obtained. We remark that the results presented here are equivalent to the algebraic solution presented in Zarzycki and Dewilde (1983a), and to the geometric solution of the stochastic estimation problem, discussed in Zarzycki (1984a,b).

The nonlinear approach to the least-squares prediction problem (for higher-order stochastic sequences) may result in better (than in the linear treatment) estimation accuracy (if the sequence is non-Gaussian), however, complexity of the generalized nonlinear filter presented here increases rapidly (synchronously with each 'global' order-update step), and becomes rather big even in relatively low-order nonlinear filters (comparing to the complexity of the linear filter), as it can be seen in Fig. 3.5. Therefore, the complexity reduction problem will be the subject of the next chapter, where time-invariant as well as 'quasi-linear' ladder-filter algorithms will be presented.

4. TIME-INVARIANT AND 'QUASI-LINEAR' LADDER-FILTERS

We noticed in the previous chapter that complexity of the generalized nonlinear ladder-filter increases rapidly (synchronously with each 'global' order-update step), and becomes relatively 'big' even in low-order nonlinear filters. Consequently, in this chapter we wish to consider the problem of complexity reduction in nonlinear ladder-filters. In order to obtain efficient nonlinear filter algorithms, we will first discuss the nonlinear least-squares prediction problem for stationary (in the higher-order sense) stochastic sequences. We will show that the solution results in the nonlinear time-invariant filter whose complexity is much reduced (comparing to the generalized algorithm). Further complexity reduction will be achieved by introducing simplified nonlinear estimation schemes, called 'quasi-linear' filters and associated with the optimum prediction of higher-order stochastic sequences whose 'distance' from the Gaussian sequence is low (in a sense to be defined). That problem has been introduced algebraically in Zarzycki and Dewilde (1983b), and considered geometrically (in the space of the Volterra functional polynomials) in Zarzycki (1984c,e).

4.1 Shift-invariance of inner-products

Let us assume that the underlying stochastic sequence $\{y\}$ is stationary (in a weak fourth-order sense). Then, following (2.19), we will obtain

$$H^{x}_{n,m} = H^{x+1}_{n,m} = H^{o}_{n,m} = T_{n,m} \tag{4.1a}$$

$$H^{x,v}_{n,m} = H^{x+1,v}_{n,m} = H^{o,v}_{n,m} = T^{v}_{n,m} \tag{4.1b}$$

regardless of the x-shift (i.e., the time-shift), where $T_{n,m}$ and $T^{v}_{n,m}$ are the generalized (block, multi-indexed) Toeplitz covariance submatrices with domains $DT_{n,m} = L^{o}_{n,m} \times L^{o}_{n,m}$ and $DT^{v}_{n,m} = L^{o,v}_{n,m} \times L^{o,v}_{n,m}$, respectively. Applying (4.1a) in (3.13), we can write

$$\begin{aligned}(F^{x}_{n,m}, G^{x}_{n,m})_{\mathbb{I}^{x}_{n,m}} &= (F^{x+1}_{n,m}, G^{x+1}_{n,m})_{\mathbb{I}^{x+1}_{n,m}} = \\ &= (F^{o}_{n,m}, G^{o}_{n,m})_{\mathbb{I}^{o}_{n,m}} = \\ &\triangleq F_{n,m} \cdot T_{n,m} \cdot \tilde{G}_{n,m} = \\ &= (F_{n,m}, G_{n,m})_{\mathbb{I}^{o}_{n,m}}\end{aligned} \tag{4.2a}$$

where

$$F_{n,m} = [f_{n,m;j_1} \quad f_{n,m;j_1,j_2}] \quad (j_1,j_2) \in L^{o}_{n,m} \tag{4.2b}$$

Applying (4.1b), we obtain similar relations for the (x,v)-labeled quantities

$$(F^{x,v}_{n,m}, G^{x,v}_{n,m})_{\mathbb{I}^{x,v}_{n,m}} \triangleq F^{v}_{n,m} \cdot T^{v}_{n,m} \cdot \tilde{G}^{v}_{n,m} = (F^{v}_{n,m}, G^{v}_{n,m})_{\mathbb{I}^{o,v}_{n,m}} \tag{4.2c}$$

with

$$F^{v}_{n,m} = [f^{v}_{n,m;j_1} \quad f^{v}_{n,m;j_1 j_2}] \quad (j_1,j_2) \in L^{o,v}_{n,m} \tag{4.2d}$$

Equations (4.2) express the x-shift (i.e., time-shift) invariance of inner-product in the higher-order (i.e., fourth-order) stationary case. In the next paragraph we will show that this property will result in significant simplifications of the nonlinear ladder-filter algorithm.

4.2 Time-invariant nonlinear ladder-filter algorithm

Following the shift-invariance of the inner-product (4.2), we notice that the L-forward approximation errors will satisfy the following relations

$$A^{x}_{n,m} = A^{x+1}_{n,m} = A^{o}_{n,m} =$$

$$\triangleq A_{n,m} = [a_{n,m;j_1} \quad a_{n,m;j_1j_2}] \quad (j_1,j_2) \in L^{o}_{n,m} \tag{4.3a}$$

For the B-forward errors (v=0,...,n), we obtain

$$A^{x,v}_{n,m} = A^{x+1,v}_{n,m} = A^{o,v}_{n,m} =$$

$$\triangleq A^{v}_{n,m} = [a^{v}_{n,m;j_1} \quad a^{v}_{n,m;j_1j_2}] \quad (j_1,j_2) \in L^{o,v}_{n,m} \tag{4.3b}$$

Similarly, for the L- and B-backward errors, we can write

$$B^{x,n-1}_{n,m} = B^{x+1,n-1}_{n,m} = B^{o,n-1}_{n,m} =$$

$$\triangleq B^{n-1}_{n,m} = [b^{n-1}_{n,m;n-j_1} \quad b^{n-1}_{n,m;n-j_1,n-j_2}] \quad (j_1,j_2) \in L^{o,n-1}_{n,m} \tag{4.4a}$$

if $m=0,\dots,n$, and for $m=n+1$

$$B^{x,n}_{n,n+1} \triangleq B^{n}_{n,n+1} = [b^{n}_{n,n+1;n-j_1} \quad b^{n}_{n,n+1;n-j_1,n-j_2}] \quad (j_1,j_2) \in L^{o,n}_{n,n+1}$$

Consequently, the forward and backward 'global' ON bases will satisfy

$$\mathbf{A}_n^x = \mathbf{A}_n^{x+1} = \mathbf{A}_n^o \triangleq \mathbf{A}_n \tag{4.5a}$$

$$\mathbf{B}_n^x = \mathbf{B}_n^{x+1} = \mathbf{B}_n^o \triangleq \mathbf{B}_n \tag{4.5b}$$

with

$$\mathbf{A}_n = [A_{n,o} \quad A_{n,1}^o \quad \cdots \quad A_{n,n+1}^n] \tag{4.5c}$$

$$\mathbf{B}_n = [B_{n,o}^{n-1} \quad \cdots \quad B_{n,n}^{n-1} \quad B_{n,n+1}^n] \tag{4.5d}$$

We notice that in stationary case, the entries of $\mathbf{B}_n$ will be used as initializations in the 'global' order-update step $n \to n+1$, yielding the higher-order forward and backward solutions $\mathbf{A}_{n+1}$ and $\mathbf{B}_{n+1}$. Consequently, in the stationary case:

a) there is no 'nesting' between the x-labeled 'levels' in the structure of the nonlinear ladder-filter;

b) the nonlinear filter algorithm can be executed at each x-labeled 'level' separately;

c) it is sufficient to run the algorithm at the (x=0)-labeled 'level' only, following (4.5).

Hence, the stationary version of the generalized nonlinear ladder-filter algorithm will be obtained if we consider the 'local' LL, LB, BL and BB recursions at the (x=0)-labeled 'level'. For example, the stationary version of the LL 'local' order-update recursion (3.42) will take the form

$$A_{n,1} = (1-[\rho_{n,1}]^2)^{-\frac{1}{2}} \; ([A_{n,o} \quad 0_{n+1}] - \rho_{n,1}[0_o \quad B_{n,o}^{n-1}]) \tag{4.6a}$$

$$B_{n,1} = (1-[\rho_{n,1}]^2)^{-\frac{1}{2}} \; (-\rho_{n,1}[A_{n,o} \quad 0_{n+1}] + [0_o \quad B_{n,o}^{n-1}]) \tag{4.6b}$$

with

$$\rho_{n,1} = ([A_{n,o} \quad 0_{n+1}] , [0_o \quad B^{n-1}_{n,o}])_{\Pi^o_{n,1}} \tag{4.6c}$$

The transform-domain counterpart of the LL recursion (4.6) will be expressed as (following (3.68))

$$\begin{bmatrix} A_{n,1}(Z) \\ \\ B_{n,1}(Z) \end{bmatrix} = \Theta_{n,1} \begin{bmatrix} A_{n,o}(Z) \\ \\ Z \cdot B^{n-1}_{n,o}(Z) \end{bmatrix} \tag{4.7a}$$

with

$$\Theta_{n,1} = (1- [\rho_{n,1}]^2)^{-\frac{1}{2}} \begin{bmatrix} 1 & -\rho_{n,1} \\ \\ -\rho_{n,1} & 1 \end{bmatrix} \tag{4.7b}$$

and

$$\rho_{n,1} = (A_{n,o}(Z) , Z \cdot B^{n-1}_{n,o}(Z))_{\mathbf{Z}} \tag{4.7c}$$

The remaining LB, BL and BB 'local' order-update recursions in the stationary case will be the 'local' recursions of Appendix 2, provided the x-label is removed.

The LL 'local' recursion (4.7) can be interpreted as the LL-section of the corresponding nonlinear ladder-filter

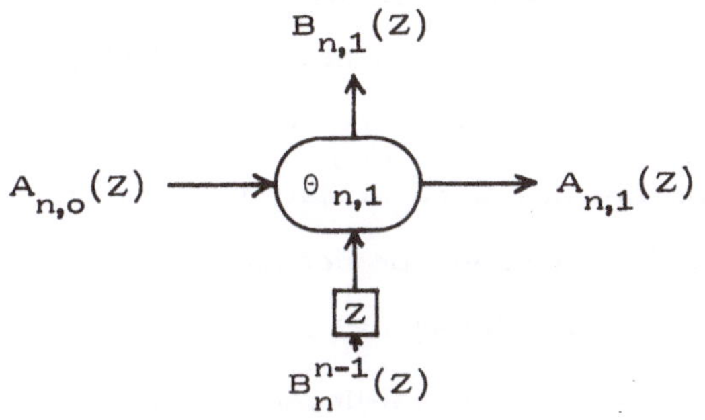

(4.7d)

The remaining 'local' LL, LB and BB sections of the stationary nonlinear filter will again be expressed as the 'local' sections of Appendix 2, provided the x-labels are removed. These sections, connected accordingly together, will constitute the 'global' section $n \to n+1$ of the filter. We notice that the set of reflection coefficients, computed by the algorithm at the 'global' step $n \to n+1$, will be expressed by (3.43), provided the x-labels are again removed. Observing that the reflection coefficients $\rho_{n,m}$ and $\overset{v}{\rho}_{n,m}$ (being actually the filter gains) do not depend on the parameter x (i.e., they do not depend on 'current' time), we conclude that the nonlinear prediction ladder-filter is time-invariant. The 'local' structure of the third-order (N=3) time-invariant nonlinear ladder-filter is presented in Fig. 4.1. We notice that the filter satisfies precisely the same orthogonality requirements, as considered in the time-variant case.

We can observe that the time-invariant nonlinear ladder-filter algorithm can be treated as the fast method for (Gram-Schmidt) orthogonalization of the basis in the space of the generalized matrices and/or of the basis in the space of the generalized z-polynomials, provided the underlying stochastic sequence is stationary (in a weak higher-order sense). On the other hand, the stationary nonlinear Levinson algorithm can be considered as the fast method for Cholesky factorization of the generalized (block, multi-indexed) Toeplitz covariance matrix. We also notice that removing all nonlinear terms, we will immediately obtain the classical linear time-invariant AR (Levinson) prediction filter for second-order stationary sequences, as considered in Dewilde, Vieira and Kailath (1978), Dewilde and Dym (1981), Kailath (1982), Deprettere and Lie (1980).

Comparing the structures of the time-variant (Fig. 3.5) and time-invariant (Fig. 4.1) nonlinear ladder-filters, we notice that significant reduction of the filter complexity has been achieved in the stationary case, although, the number of the quantities processed in the filter will still increase synchronously with each 'global' order-update step. Further complexity reduction will be achieved in 'quasi-linear' prediction filters.

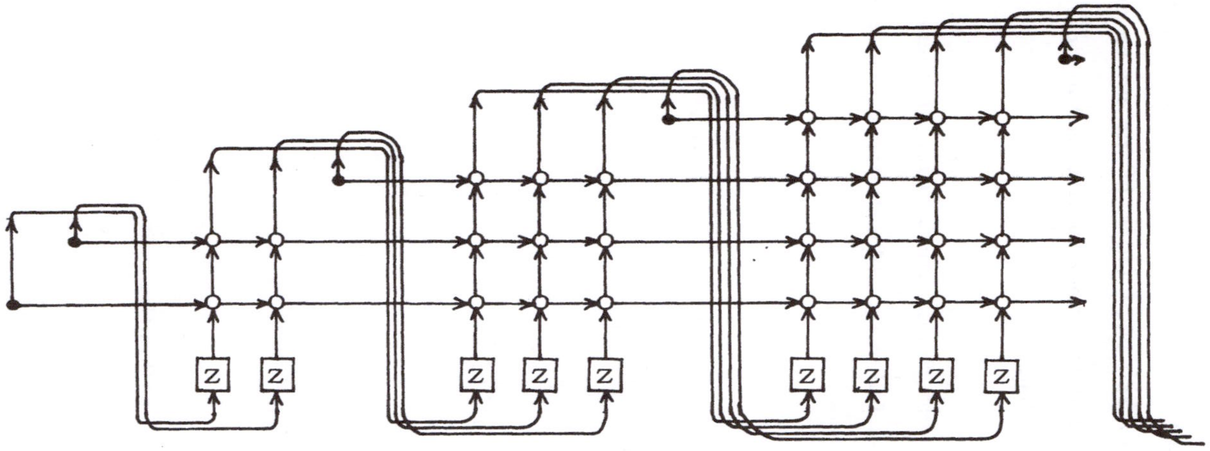

Fig. 4.1 'Local' structure of the third-order (N=3) time-invariant nonlinear ladder-filter. All symbols have the same meaning as in Fig. 3.5.

4.3 'Quasi-linear' ladder-filters

In this paragraph we will consider a class of simplified nonlinear ladder-filter algorithms which we will call 'quasi-linear' filters. These filters will yields better estimation accuracy (than in the linear case) while their complexity will be reduced in comparison with the previously considered nonlinear filter algorithms.

Let us assume that the underlying fourth-order stochastic sequence $\{y\}$ is represented by the following submatrices of the random variables (and their products)

$$Y^{1,n}_{n,n+1} = \left[\begin{array}{c} y_{-j_1} \\ \hline y_{-j_1} y_{-j_2} \end{array} \right]_{(j_1,j_2) \in L^{1,n}_{n,n+1}} , \quad n=0,\ldots,N-1 \tag{4.8a}$$

where

$$L^{1,n}_{n,n+1} = L^1_n \cup {}^2L^{1,n}_{n,n+1} \tag{4.8b}$$

with $L^1_n = \{1,\ldots,n+1\}$ and

$${}^2L^{1,n}_{n,n+1} = \mathrm{sym}\,{}^2L^1_n \times \mathrm{sym}\,{}^2L^1_n \tag{4.8c}$$

Now let us introduce for $n=0,\ldots,N-1$ and $\beta=0,\ldots,n+1$ the following index-sets

$$L^{(\beta)}_n \triangleq L^1_n \cup {}^2L^{(\beta)}_n \tag{4.9a}$$

where the bi-variate part of the index-set (4.9a) is given by

$$^{2}L_n^{(\beta)} \triangleq \{(j_1,j_2): j_1 \in L_n^1 \ , \ j_2=j_1,\dots,\beta\} \tag{4.9b}$$

Then we can observe that:

- if $\beta=0$ then $L_n^{(o)} = L_n^1$ (since $^{2}L_n^{(o)} = \phi$);
- if $\beta=n+1$ then $L_n^{(n+1)} = L_{n,n+1}^{1,n}$ (since $^{2}L_n^{(n+1)} = {}^{2}L_{n,n+1}^{1,n}$);
- if $0 < \beta < n+1$ then $L_n^{(o)} \subset L_n^{(\beta)} \subset L_n^{(n+1)}$.

If we introduce the index-set

$$^{e}L_n^{(\beta)} \triangleq {}^{2}L_{n,n+1}^{1,n} \setminus {}^{2}L_n^{(\beta)} =$$

$$= \{(k_1,k_2): k_1=\beta+1,\dots,n \ , \ k_2=k_1,\dots,n\} \tag{4.10}$$

then we notice that:

- if $\beta=0$ then $^{e}L_n^{(o)} = {}^{2}L_{n,n+1}^{1,n}$;
- if $\beta=n+1$ then $^{e}L_n^{(n+1)} = \phi$;
- if $0 < \beta < n+1$ then $^{e}L_n^{(\beta)} \subset {}^{2}L_{n,n+1}^{1,n}$.

Following (4.9), we can consider for $n=0,\dots,N-1$ and $\beta=0,\dots,n+1$ the submatrices

$$Y_n^{(\beta)} \triangleq \begin{bmatrix} y_{-j_1} \\ \hline y_{-j_1}y_{-j_2} \end{bmatrix}_{(j_1,j_2) \in L_n^{(\beta)}} \tag{4.11}$$

and we mention that:

- if $\beta=0$ then $Y_n^{(o)} = [\, y_{-j_1} \,]_{j_1 \in L_n^1}$;
- if $\beta=n+1$ then $Y_n^{(n+1)} = Y_{n,n+1}^{1,n}$.

Following (4.10), we will introduce the submatrices

$$^{e}Y_n^{(\beta)} \triangleq [y_{-j_1}y_{-j_2}]_{(j_1,j_2) \in {}^{e}L_n^{(\beta)}} \tag{4.12}$$

If the fourth-order stochastic sequence $\{y\}$ was Gaussian, then (2.10) with $M=2$ would hold. Now let us suppose that (2.10) applies for the submatrix ${}^{\theta}Y_n^{(\beta)}$ only; i.e., we have

$$\mathbb{E}\{y_o y_{-k_1} y_{-k_2}\} = 0 \qquad \text{for} \qquad (k_1,k_2) \in {}^{\theta}L_n^{(\beta)} \tag{4.13}$$

This means that the sequence $\{y\}$ is 'partially' Gaussian, or, ' β-Gaussian', and its Gaussian part is determined by the submatrix ${}^{\theta}Y_n^{(\beta)}$.
From (4.9) and (4.10) it follows that:

- the sequence is just Gaussian if $\beta=0$ (since ${}^{\theta}Y_n^{(o)} = {}^{2}Y_{n,n+1}^{1,n}$);
- the sequence is non-Gaussian if $\beta=n+1$ (since $D^{\theta}Y_n^{(n+1)} = \phi$);
- the sequence is ' β-Gaussian' if $0 < \beta < n+1$, with

$${}^{2}Y_n^{(\beta)} = [y_{-j_1} y_{-j_2}] \quad (j_1,j_2) \in {}^{2}L_n^{(\beta)} \tag{4.14}$$

indicating the non-Gaussian part of that sequence, and with ${}^{\theta}Y_n^{(\beta)}$ being its Gaussian part.

If the value of the parameter β is low, we will say that the sequence is 'quasi-Gaussian'.

Following (4,8)-(4.14), we can consider (under the isomorphism (2.47)) the subspaces (for $n=0,\ldots,N-1$ and $\beta=0,\ldots,n+1$)

$$\mathbb{I}_{n,n+1}^{1,n} = \vee \{I_{n,n+1}^{1,n}\} \tag{4.15a}$$

$$\mathbb{I}_n^{(\beta)} \triangleq \vee \{I_n^{(\beta)}\} \tag{4.15b}$$

$${}^{\theta}\mathbb{I}_n^{(\beta)} \triangleq \vee \{{}^{\theta}I_n^{(\beta)}\} \tag{4.15c}$$

where

$$I_n^{(\beta)} \triangleq [1_{j_1} \quad 1_{j_1 j_2}] \quad (j_1,j_2) \in L_n^{(\beta)} \tag{4.15d}$$

$$ {}^{e}I_n^{(\beta)} \triangleq [1_{k_1k_2}]_{(k_1,k_2)\in {}^{e}L_n^{(\beta)}} \tag{4.15e} $$

Then we can observe that

$$ I_{n,n+1}^{1,n} = II_n^{(\beta)} \oplus {}^{e}I_n^{(\beta)} \tag{4.16} $$

where $\oplus$ stands for direct (although not orthogonal) sum of subspaces. Using (4.15),(4.16), and according to (2.47), the orthogonality conditions (4.13) can be rewritten as

$$ 1_o \perp {}^{e}I_n^{(\beta)} \tag{4.17} $$

Now let us define the following approximation error

$$ A_{n,o}^{(\beta)} \triangleq P_{II;n}^{(\beta)} 1_o \perp II_n^{(\beta)} \tag{4.18a} $$

and let $A_{n,o}^{(\beta)}$ denote its normalized version; i.e.,

$$ A_{n,o}^{(\beta)} = A_{n,o}^{(\beta)} \| A_{n,o}^{(\beta)} \|_{II}^{-1} \tag{4.18b} $$

(where $P_{II;n}^{(\beta)}$ is the orthogonal projection operator on the subspace $II_n^{(\beta)}$). Then we can show that

$$ A_{n,o}^{(\beta)} \perp {}^{e}I_n^{(\beta)} \tag{4.19} $$

Indeed, let

$$ \hat{1}_{n,o}^{(\beta)} \triangleq P_{II;n}^{(\beta)} 1_o $$

From (4.17) it follows that

$$\hat{1}^{(\beta)}_{n,o} = P^{(\beta)}_{\mathbb{I};n}\, 1_o = P^{1,n}_{\mathbb{I};n,n+1}\, 1_o$$

This implies

$$A^{(\beta)}_{n,o} = 1_o - \hat{1}^{(\beta)}_{n,o} \perp \mathbb{I}^{1,n}_{n,n+1}$$

and, consequently,

$$A^{(\beta)}_{n,o} \perp {}^{\ominus}\mathbb{I}^{(\beta)}_{n}$$

as ${}^{\ominus}\mathbb{I}^{(\beta)}_{n} \subset \mathbb{I}^{1,n}_{n,n+1}$. Hence, for $(k_1,k_2) \in {}^{\ominus}L^{(\beta)}_{n}$,

$$A^{(\beta)}_{n,o} \perp 1_{k_1 k_2}$$

This means that the use of the subspace ${}^{\ominus}\mathbb{I}^{(\beta)}_{n}$ will not imply better estimation accuracy, and it is useless to include that subspace into the nonlinear estimation scheme for the underlying 'β-Gaussian' sequence. Consequently, the optimum estimation scheme for that sequence will be associated with the subspace $\mathbb{I}^{(\beta)}_{N}$ (expressing the non-Gaussian part of the sequence, under the isomorphism (2.47)). We can observe that in the Gaussian case (corresponding to $\beta = 0$ and, hence, to the Gaussian part ${}^{\ominus}\mathbb{I}^{(0)}_{N} = {}^{2}\mathbb{I}^{1,N-1}_{N-1,N}$) it is sufficient to consider the 'uni-variate' part $\mathbb{I}^{1}_{N-1} = \vee\{1_{j_1}, j_1 = 1,\dots,N\}$ in the optimum estimation scheme. This means that the best filter for a Gaussian sequence is the linear ladder-filter (being actually the 'most simple nonlinear filter'). Therefore, it can be expected that the optimum nonlinear ladder-filter, associated with a 'β-Gaussian' sequence (where $0 < \beta < n+1$) should be less complex than the general nonlinear filter, while estimation accuracy should be still better than in the linear treatment.

In order to show that, it is convenient to introduce a notion of the 'nonlinear rank' of the filter, as the number of the v-labeled 'levels', existing in the filter structure after N 'global' order-update steps. We can observe (see Zarzycki (1984c)) that the optimum nonlinear filter for a 'β-Gaussian' sequence is the filter whose nonlinear rank equals β. This filter will be called the 'β-linear' filter. If the value of the parameter β is low, we shall say that the filter is 'quasi-linear'. Then we can observe that:

- if $\beta=0$ then we obtain the linear ladder-filter algorithm (of the smallest complexity);
- if $\beta=N$ then we get the general time-invariant nonlinear ladder-filter algorithm, introduced in the previous paragraph (of biggest complexity);
- if $0<\beta<N$ then the filter complexity will be bigger than in the linear algorithm but smaller than in the general nonlinear case, since the ordering in the 'β-linear' filter algorithm is

$$\begin{array}{lll} \text{modulo } (n+2) & \text{if} & n=0,\ldots,\beta-1 \\ \text{modulo } (\beta+1) & \text{if} & n=\beta,\ldots,N \end{array} \qquad (4.20)$$

This means that the number of the v-labeled 'levels' in the filter structure is growing from 1 up to β synchronously with each 'global' order-update step (for $n=0,\ldots,\beta-1$) since at each step the one 'new' B-forward and B-backward error is introduced to the scheme, much like in the general nonlinear filter algorithm. Then, starting with $n=\beta$, the number of the v-labeled 'levels' is kept constant (and equals β), regardless of the 'length' of the filter. Thus, for $n=\beta,\ldots,N$ the filter algorithm computes the $(\beta+1)^2$ reflection coefficients per 'global' order-update. Therefore, the filter complexity is considerably reduced, in comparison with the general ($\beta=N$) scheme. This is illustrated in Fig. 4.2 where the 'quasi-linear' filters ($\beta=0,1,2,3$) are presented.

Now let us evaluate estimation accuracy in the 'β-linear' filters. Following (3.48), we can observe that the error-norm relations in the ge-

D)

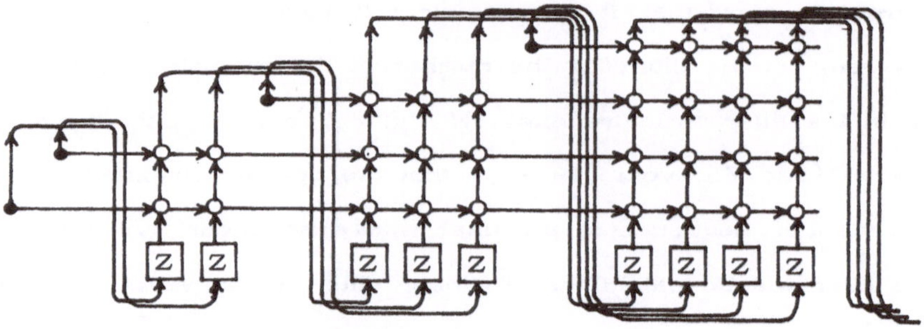

C)

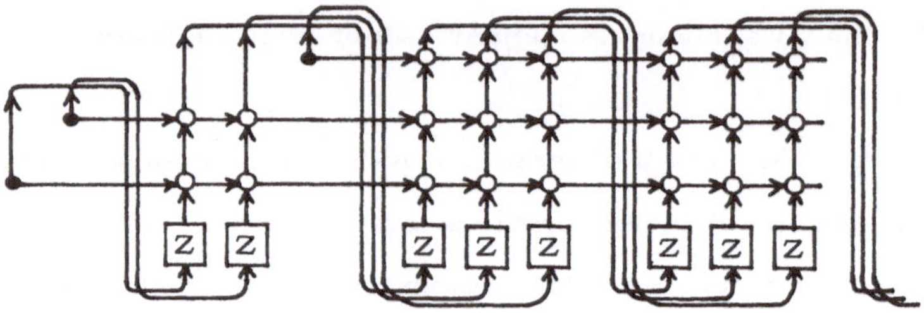

B)

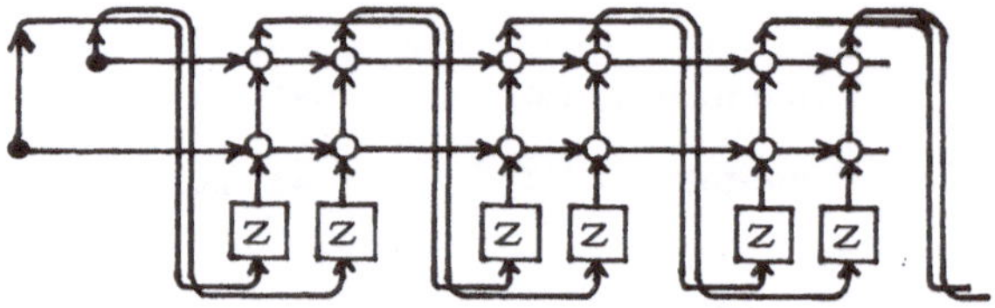

A)

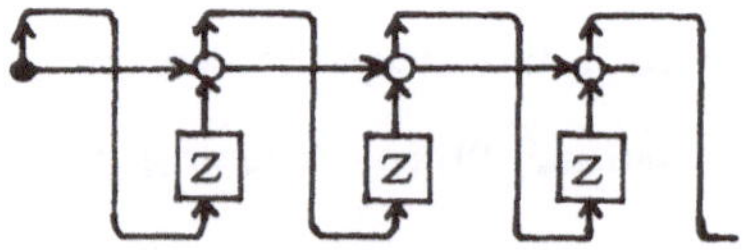

Fig. 4.2 'Quasi-linear' ladder-filters.

A) $\beta = 0$
B) $\beta = 1$
C) $\beta = 2$
D) $\beta = 3$

neral time-invariant nonlinear ladder-filter are given by

$$\| A_{N,o} \| = \| A_{o,o} \| \, R_{LL} \, R_{LB} \tag{4.21a}$$

where

$$R_{LL} = \prod_{n=0}^{N-1} (1 - [\rho_{n,1}]^2)^{\frac{1}{2}} \tag{4.21b}$$

$$R_{LB} = \prod_{n=0}^{N-1} \left[\prod_{m=2}^{n+1} (1 - [\rho_{n,m}]^2)^{\frac{1}{2}} \right] (1 - [\rho_{n+1,o}]^2)^{\frac{1}{2}} \tag{4.21c}$$

Then we can observe that in the ' β-linear' scheme

$$\| A_{N,o}^{(\beta)} \| = \| A_{o,o} \| \, R_{LL} \, R_{LB}^{(\beta)} \tag{4.22a}$$

where

$$R_{LB}^{(\beta)} = R_1^{(\beta)} \, R_2^{(\beta)} \tag{4.22b}$$

with

$$R_1^{(\beta)} = \prod_{n=0}^{\beta-1} \left[\prod_{m=2}^{n+1} (1 - [\rho_{n,m}^{(\beta)}]^2)^{\frac{1}{2}} \right] (1 - [\rho_{n+1,o}^{(\beta)}]^2)^{\frac{1}{2}} \tag{4.22c}$$

$$R_2^{(\beta)} = \prod_{n=\beta}^{N-1} \left[\prod_{m=2}^{\beta} (1 - [\rho_{n,m}^{(\beta)}]^2)^{\frac{1}{2}} \right] (1 - [\rho_{n+1,o}^{(\beta)}]^2)^{\frac{1}{2}} \tag{4.22d}$$

We notice that the norm of the error in the 'β -linear' case is reduced (with respect to the linear case) by the factor $R_{LB}^{(\beta)}$ (4.22b), and we can observe that:

- if $\beta=0$ then $R_{LL} R_{LB}^{(o)} = R_L = \prod_{n=0}^{N-1} (1 - [\rho_{n+1}]^2)^{\frac{1}{2}}$
- if $\beta=N$ then $R_{LL} R_{LB}^{(N)} = R_{LL} R_{LB}$ (with R_{LB} given by (4.21c)).

Consequently, we can associate with each ' β-Gaussian' sequence the optimum ' β-linear' filter and, working with not too complex nonlinear orthogonal structures (whose complexity may be successively increased until the desired estimation accuracy is achieved), we will obtain better accuracy than in the linear treatment.

4.4 Experimental example

The 'β-linear' ladder-filters have been tested using pseudo-Gaussian and non-Gaussian excitations. In Figs. 4.3 - 4.6 we present computer plots of the mean-square-errors (MSE) in the adaptive 'β-linear' (β=0,1,2,3) innovations filters of the eight-order (N=8), associated with 20 ms samples of the input Gaussian and non-Gaussian time-series.

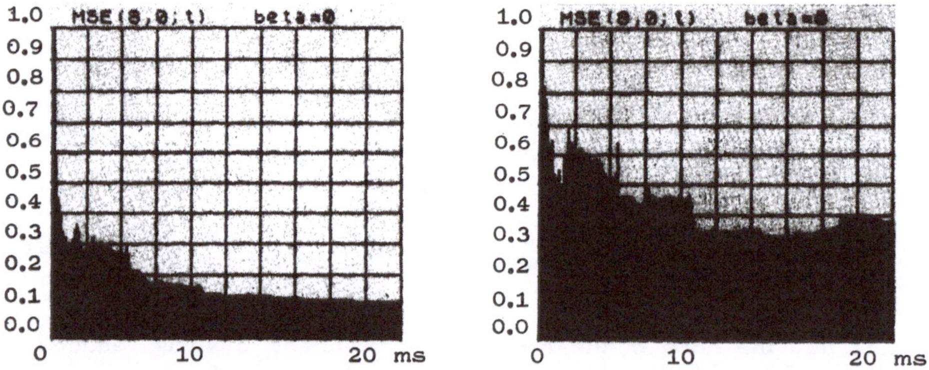

A) B)

Fig. 4.3 Mean-square error in the '0-linear', 8-th order ladder-filter, inputted with:
A) Gaussian
B) non-Gaussian
excitations.

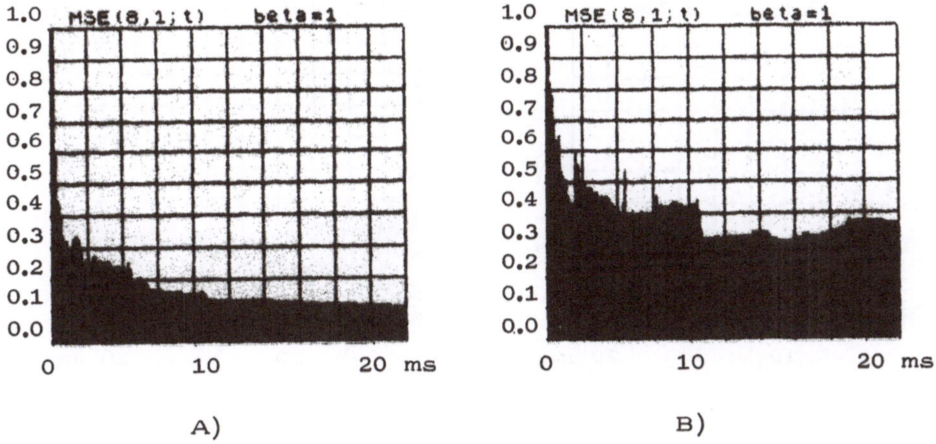

Fig. 4.4 Mean-square-error in the '1-linear', 8-th order ladder-filter, inputted with:
A) Gaussian
B) non-Gaussian
excitation.

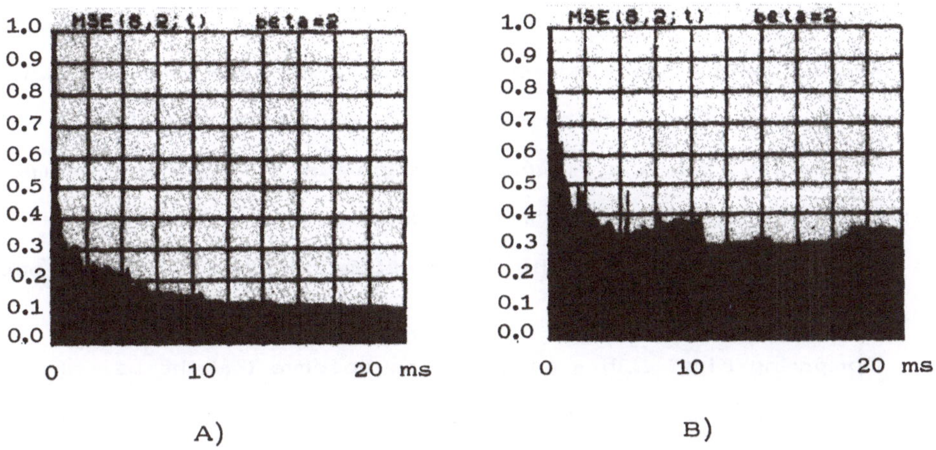

Fig. 4.5 Mean-square-error in the '2-linear', 8-th order ladder-filter, inputted with:
A) Gaussian
B) non-Gaussian
excitations.

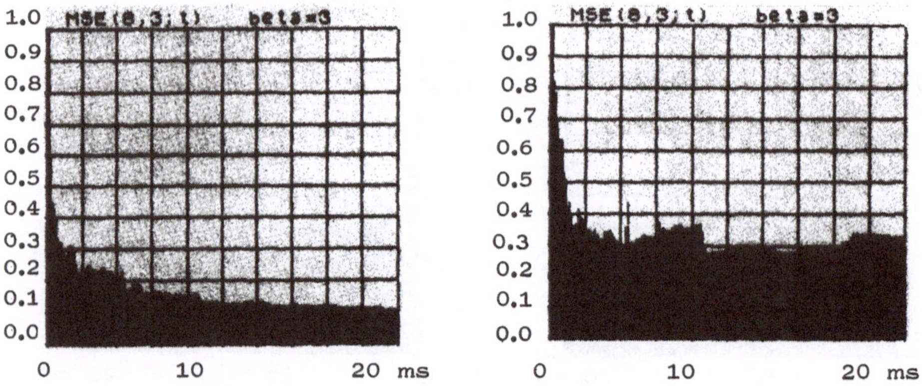

A) B)

Fig. 4.6 Mean-square-error in the '3-linear', 8-th order ladder-filter, inputted with:
A) Gaussian
B) non-Gaussian
excitations.

Comparing Figs. 4.3a and 4.3b, we can observe that the linear estimation accuracy in the non-Gaussian case is much worse than in case of the Gaussian excitation. This follows from the fact that the linear ladder-filter operates on the second-order statistics of the input time-series, which are not sufficient in order to characterize non-Gaussian signals.

Comparing Figs. 4.3a - 4.6a, we notice that the nonlinear estimation schemes do not imply better estimation accuracy in the Gaussian case. This results from the fact that the linear estimation filter is the best possible filter for a Gaussian signal.

Comparing Figs. 4.3b - 4.6b, we can observe that the use of (even the most simple) nonlinear estimation procedures introduce significant improvement of estimation accuracy in case of non-Gaussian excitation.

It should be noted that the improvement of estimation accuracy actually depends on the higher-order statistics of the underlying sequence, however, that improvement can be achieved if a (suitably chosen) nonlinear filter is used.

5. CONCLUDING REMARKS

The nonlinear prediction filter algorithms, presented in this work, can be directly implemented in a soft- and/or in a hard-ware way. The modular structure of the nonlinear orthogonal ladder-filters implies soft-ware realizations, requiring relatively small capacity of the operational memory. A hard-ware realization follows from the fact that the basic orthogonal 'building-block' of the nonlinear filter (actually the Given's rotor) is precisely the same as in the linear ladder-filters. That 'building-block' can be implemented using VLSI integrated circuits (namely CORDICS processors), introduced in the realizations of the linear orthogonal filters (see e.g., Ahmed and Morf (1982); Deprettere (1983b); Deprettere, Dewilde and Udo (1984); Dewilde (1983)). Consequently, the nonlinear digital filters, considered here, can be realized with those processors as well, taking advantage of the parallel computations. It should be noted that the use of the normalized orthogonal 'building-blocks' assures inherent numerical stability of the nonlinear filters. Moreover, adaptive versions of those nonlinear prediction filters, operating directly on a stream of data, and having parameter-tracking capability, can also be introduced (see Zarzycki (1984d)). The nonlinear adaptive filter algorithms result in the 'exact nonlinear least-squares' solution at each time-instant, much like in the linear case considered by Lee, Morf and Friedlander (1981). Consequently, the nonlinear prediction filter algorithms, introduced in this work, are also suitable for on-line nonlinear processing of higher-order time-series.

REFERENCES

AHMED H.M. and MORF M.

1982 VLSI array architectures for matrix factorization, in Outils et modeles mathematiques pour l'automatique, l'analyse de systemes et le traitement du signal, Ed.CNRS, Paris, vol.2, pp. 691-704.

BARRET J.F.

1963 The use of functionals in the analysis of nonlinear physical systems, J.Electr.Contr., vol.15, pp.567-615.

BOSE A.G.

1958 A theory of nonlinear systems, Techn.Rept., MIT.

CAMERON R.H. and MARTIN W.T

1947 The orthogonal development of nonlinear functionals in series of Fourier-Hermite functionals, Ann.Math., vol.48, pp. 385-392.

DELSARTE P., GENIN Y. and KAMP Y.

1979 a Schur parametrization of positive definite block-Toeplitz matrices, SIAM J.Appl.Math., vol.36, pp.33-46.

1979 b The Nevanlinna-Pick problem for matrix-valued functions, SIAM J.Appl.Math., vol.36, pp.47-61.

1983 Generalized Schur positivity test and Levinson recursion, Proc.ECCTD'83, Stuttgart.

DELOSME J.M. and MORF M.

1982 Fast algorithms for finite shift-rank processes : geometric approach, in Outils et modeles mathematiques pour l'automatique, l'analyse de systemes et le traitement du signal, Ed.CNRS, Paris, vol.2, pp.499-529.

DEPRETTERE E.

1981 Orthogonal filters, Ph.D. Thesis, Delft Univ. Techn.

1982 Mixed form time-variant lattice recursions, in Outils et modeles mathematiques pour l'automatique, l'analyse de systemes et le traitement du signal, Ed.CNRS, Paris, vol.2, pp.545-562.

1983 a Synthesis and fixed-point implementation of pipelined true orthogonal filters, Proc.ICASSP'83, Boston.

1983 b CORDIC-10: An expandable VLSI implementable orthogonal filter module, Proc.EUSIPCO'83, Erlangen.

DEPRETTERE E. and DEWILDE P.

1979 Generalized orthogonal filters for stochastic prediction and modeling, in Digital signal processing, Ed.V.Capellini, Acad. Press, N.Y.

DEPRETTERE E., DEWILDE P., and UDO R.

1984 Pipelined CORDIC architectures for fast VLSI filtering and array processing, Proc.ICASSP'84.

DEPRETTERE E. and JAINANDUNSING K.

1984 Design and VLSI implementation of a concurrent solver for N coupled systems of linear equations, Techn.Rept., Delft Univ. Techn.

DEPRETTERE E. and LIE S.C.

1980 Generalized Schur-Darlington algorithms for lattice-structured matrix inversion and stochastic modeling, Techn.Rept., Delft Univ. Techn.

DEWILDE P.

1982 Stochastic modeling with orthogonal filters, in Outils et modeles mathematiques pour l'automatique, l'analyse de systemes et le traitement du signal, Ed.CNRS, Paris, vol.2, pp.331-398.

1983 Orthogonal filters: Pipelining and VLSI implementation, Proc. ECCTD'83, Stuttgart.

1984a Spectral approximation and estimation with scattering functions, in Mathematical Theory of Networks and Systems, Lecture Notes in Control and Information Sciences, vol.48, Ed.P.A.Fuhrmann, Springer-Verlag, pp.234-252.

1984b Orthogonal filters: A numerical approach to filtering theory, Ibid., pp.253-267.

DEWILDE P. and BULTHEEL A.

1979 Orthogonal functions related to the Nevanlinna-Pick problem, in Mathematical Theory of Networks and Systems, Ed.P.Dewilde, vol.3, Delft, pp.207-212.

DEWILDE P., DEPRETTERE E. and NOUTA R.

1984 Parallel and pipelined VLSI implementations of signal processing algorithms, in VLSI and signal processing, Ed.S.Y. Kung.

DEWILDE P. and DYM H.

1981a Schur recursions, error formulas and convergence of rational estimator for stationary stochastic processes, IEEE Trans. on IT-27, pp.446-461.

1981b Lossless chain scattering matrices and optimum linear prediction: The vector case, Circuit Theory and Appl., vol.9, pp.135-175.

1984 Lossless inverse scattering with rational networks: Theory and applications, IEEE Trans. on IT-30.

DEWILDE P., VIEIRA A.C. and KAILATH T.

1978 On a generalized Szegö-Levinson realization algorithm for optimal linear predictors based on a network synthesis approach, IEEE Trans. on CAS-25, pp.663-675.

FRECHET M.

1910 Sur les fonctionelles continues, Ann. de l'Ecole Norm. Sup. 3-me, Ser.V.27.

ITO K.

1951 Multiple Wiener Integral, J.Math.Soc., Japan, vol.13, nr 1, pp.157-169.

KAILATH T.

1974 A view of three decades in linear filtering theory, IEEE Trans. on IT-20, pp.146-181.

1982 Time-variant and time-invariant lattice filters for nonstationary processes, in Outils et modeles mathematiques pour l'automatique, l'analyse de systemes et le traitement du signal, Ed.CNRS, Paris, vol.2, pp.417-464.

LEE D.T.L., MORF M. and FRIEDLANDER B.

1981 Recursive least-squares ladder-estimation algorithms, IEEE Trans. on CAS-28, pp.467-481.

LEV-ARI H.

1982 Parametrization and modeling of nonstationary processes, Ph.D.Thesis, Stanford Univ.

1983 Modular architectures for adaptive multichannel lattice algorithms, Proc.ICASSP'83.

LEV-ARI H. and KAILATH T.

1982 Lattice filter parametrization and modeling of nonstationary processes, Techn.Rept., Stanford Univ.

LEVINSON N.

1947 The Wiener RMS error criterium in filter design and prediction, J.Math.Phys., vol.25, pp.261-278.

MORF M., VIEIRA A.C., LEE D.T.L. and KAILATH T.

1978 Recursive multichannel maximum entropy spectral estimation, IEEE Trans. on GE-16, pp.85-94.

OGURA H.

1972 Orthogonal functionals for the Poisson process, IEEE Trans. on IT-18, pp.473-481.

PIEKARSKI M.S.

1971 Reciprocal Darlington section suitable for an integrated circuit, Electron.Lett., vol.7, pp.475-477.

1974 A minimal grounded cascade synthesis for integrated circuits, Proc.ECCTD'74, London.

PIEKARSKI M.S., SAEED K.

1980 A test for positive real function, Proc.ECCTD'80, Warsaw.

PIEKARSKI M.S. and URUSKI M.

1984 Interpolation with positive real matrices, Proc.ISYNT'84, Sarajevo.

PRABHAKARA RAO C.V.K. and HELMOND J.

1983 On the theory of AR spectral approximation for processes containing deterministic signals, Proc.ECCTD'83, Stuttgart.

SEGALL A. and KAILATH T.

1976 Orthogonal functionals of independent-increment processes, IEEE Trans. on IT-22, pp.287-298.

SCHETZEN M.

1980 Volterra-Wiener theories of nonlinear systems, Wiley, N.Y.

SCHUR J.

1917 Uber potenzreichen, die in innern des einheitskreises beschrankt sind, J.Reine Ang.Math., vol.147, pp.205-232.

STEINHAUS H. and KACZMARZ S.

1935 Theorie der orthogonalreihen, Warsaw.

TUSZYNSKI A.A.

1980 A CORDIC arithmetic processor chip, IEEE Trans. on C-29, pp.68-79.

VICTOR J. and KNIGHT B.

1979 Nonlinear analysis with arbitrary stimulus ensemble, Quart.Appl.Math., vol.XXXVII, pp.115-136.

VOLTERRA V.

1959 Theory of functionals and of integral and integro-differential equations, Dover Publ.

WIDYA I.

1982 Continuous-time stochastic modeling with lossless structures, Ph.D.Thesis, Delft Univ.Techn.

WIENER N.

1938 The homogeous chaos, Amer. J.Math., vol.60, pp.897-936.

1958 Nonlinear problems in random theory, MIT Press - Wiley N.Y.

WOLDER P.

1959 The CORDIC trigonometric computing technique, IRE Trans. on EC-8, pp.330-334.

YASUI S.

1979 Stochastic functional Fourier series, Volterra series and nonlinear system analysis, IEEE Trans. on AC-21, pp. 230-242.

ZARZYCKI J.

1983 Nonlinear Levinson prediction filter for higher-order random sequences, Proc.ECCTD'83, Stuttgart.

1984a Nonlinear prediction of higher-order random sequences, submitted for publication.

1984b Generalized ladder-filters for nonlinear prediction of higher-order random sequences, submitted for publication.

1984c Fast algorithms for the least-squares nonlinear prediction, submitted for publication.

1984d Adaptive properties of nonlinear ladder-filters, submitted for publication

1984e Nonlinear ladder-filters for the least-squares AR prediction of higher-order random sequences, Proc.ISCAS'84, Montreal.

1985a Nonlinear Levinson algorithm: A geometric approach, Proc.ECCTD'85, Prague.

1985b Orthogonal ladder-form representations of nonlinear prediction filters of the Volterra-Wiener class, in Mathematical Theory of Networks and Systems, to be published.

ZARZYCKI J. and DEWILDE P.

1983a Nonlinear least-squares prediction of higher-order random sequences, submitted for publication.

1983b The Levinson-type filters for fast nonlinear AR prediction, Techn.Rept., Wroclaw Univ.Techn.

APPENDIX 1

MULTI-INDEXED MATRICES AND GENERALIZED MATRIX THEORY

Let I denote a contiguous subset of integers, let $\mathbb{R}$ be the set of real numbers. We define a m-indexed matrix ${}^{m}A$ as a map

$$ {}^{m}A : {}^{m}I \rightarrow \mathbb{R} \tag{A.1} $$

where ${}^{m}I = I \times \dots \times I$ (m-copies). According to (A.1), the index-set ${}^{m}I$ can be called the domain of the m-indexed matrix ${}^{m}A$. This domain will be denoted as $D\,{}^{m}A$. Let us introduce the index-sets

$$ L_n^o \triangleq \{j\}_o^n = \{0,1,\dots,n\} \tag{A.2} $$

$$ {}^{m}L_n^o \triangleq \underbrace{L_n^o \times \dots \times L_n^o}_{m} = \{(j_1,\dots,j_m) : j_k \in L_n^o ,\ k=1,\dots,m\} \tag{A.3} $$

A m-indexed matrix ${}^{m}\underline{A}_n$ will be called the n-th order matrix if

$$ D\,{}^{m}\underline{A}_n = {}^{m}L_n^o \tag{A.4a} $$

This matrix can be equivalently expressed in terms of its m-indexed entries as follows

$$ {}^{m}\underline{A}_n = [\, a_{j_1 \dots j_m} \,]_{(j_1,\dots,j_m) \in {}^{m}L_n^o} \tag{A.4b} $$

We will consider here some properties of multi-indexed matrices, and introduce operations on those matrices. We will usually drop, for simplicity, the order and domain of the matrices, assuming that all matrices are of the type (A.4), unless otherwise stated.

Symmetric matrix

A m-indexed matrix will be called symmetric if for any permutation $\pi_1,\dots,\pi_m$ of integers $1,\dots,m$ we shall have

$$\underline{a}_{j_{\pi_1}\dots j_{\pi_m}} = \underline{a}_{j_1\dots j_m} \qquad \text{(A.5a)}$$

Consequently, it is sufficient to consider $\binom{m+n}{m}$ 'different' entries of the symmetric matrix instead of n^m entries of non-symmetric matrix. Now let $\gamma_{j_1\dots j_m}$ denote the number of equal elements of the symmetric matrix, corresponding to the sequence $(j_1,\dots,j_m)$. We shall denote the 'symmetric part' of a m-indexed matrix ${}^m\underline{A}_n$ by ${}^mA$, where

$${}^mA_n = [\, a_{j_1\dots j_m} \,]_{(j_1,\dots,j_m)\,\in\,\mathrm{sym}\,{}^mL_n^o} \qquad \text{(A.5b)}$$

with $\mathrm{sym}\,{}^mL_n^o$ denoting the 'symmetric part' of the m-variate index-set, obtained according to lexicographic or anti-lexicographic ordering. The entries $a_{j_1\dots j_m}$ of ${}^mA_n$ can then be expressed in terms of the entries $\underline{a}_{j_1\dots j_m}$ of the symmetric matrix ${}^m\underline{A}_n$ as

$$a_{j_1\dots j_m} = \gamma_{j_1\dots j_m}\, \underline{a}_{j_1\dots j_m} \qquad \text{(A.5c)}$$

Transpose matrix

Let π be a permutation of the index-set $\{(j_1,\dots,j_m) \in {}^mL_n^o\}$. π may be represented by a map

$$\begin{pmatrix} 1, & 2, & \dots, & m \\ \pi_1, & \pi_2, & \dots, & \pi_m \end{pmatrix} \qquad \text{(A.6a)}$$

where $(\pi_1, \pi_2, \ldots, \pi_m)$ is a permutation of $\{1,2,\ldots,m\}$. Then, a m-indexed matrix ${}^m A'_\pi$ will be called the transpose matrix due to the permutation π if

$$({}^m A'_\pi)_{j_1 \ldots j_m} = ({}^m A)_{\pi (j_1, \ldots, j_m)} \tag{A.6b}$$

Zero-matrix

A m-indexed matrix will be called a zero-matrix if for each sequence of indices $(j_1, \ldots, j_m) \in D\, {}^m A$ we have $a_{j_1 \ldots j_m} = 0$. This matrix will be denoted by

$${}^m O_n = [0_{j_1 \ldots j_m}] \quad (j_1, \ldots, j_m) \in D\, {}^m O_n \tag{A.7}$$

where $0_{j_1 \ldots j_m}$ will be the zero-entry with 'coordinates' $(j_1, \ldots, j_m)$.

Unit-matrix

A 2m-indexed matrix will be called the unit-matrix if for each $(j_1, \ldots, j_m)$ and $(k_1, \ldots, k_m) \in D\, {}^{2m} A_n$ we have

$$a_{j_1 \ldots j_m k_1 \ldots k_m} = \delta_{j_1 \ldots j_m ; k_1 \ldots k_m} \tag{A.8a}$$

where

$$\delta_{j_1 \ldots j_m ; k_1 \ldots k_m} = \begin{cases} 1 & \text{if } j_1 = k_1, \ldots, j_m = k_m \\ 0 & \text{otherwise} \end{cases} \tag{A.8b}$$

This matrix will be denoted as

$$ {}^{2m}\mathbf{1}_n = [\delta_{j_1 \dots j_m ; k_1 \dots k_m}]_{(j_1, \dots, j_m, k_1, \dots, k_m) \in D^{2m}\mathbf{1}_n} \tag{A.8c} $$

<u>Block-matrices</u>

A block-matrix whose block-entries are m-indexed, n-th order matrices

$$ {}^{\{M\}}A_n = [{}^{m}A_n]_{m=1,\dots,M} \tag{A.9a} $$

will be called a M-block (row), m-indexed, n-th order matrix. Its block-row domain $D\,{}^{\{M\}}A_n$ will be a vector of simple domains

$$ D\,{}^{\{M\}}A_n = [D\,{}^{m}A_n]_{m=1,\dots,M} \tag{A.9b} $$

Similarly, a block-matrix

$$ {}^{\{M\}}B_n = \text{col}\,[{}^{m}B_n]_{m=1,\dots,M} \tag{A.10a} $$

will be called a M-block (column), m-indexed, n-th order matrix with the block-column domain

$$ D\,{}^{\{M\}}B_n = \text{col}\,[D\,{}^{m}B_n]_{m=1,\dots,M} \tag{A.10b} $$

Finally, a block-matrix

$$ {}^{\{M\times M\}}H_n = [{}^{m\oplus u}H_n]_{m,u=1,\dots,M} \tag{A.11a} $$

whose block-entries are (m+u)-indexed matrices

$$ {}^{m\oplus u}H_n = [h_{j_1 \dots j_m k_1 \dots k_u}]_{(j_1,\dots,j_m,k_1,\dots,k_u) \in D^{m\oplus u}H_n} \tag{A.11b} $$

will be called a $(M{\times}M)$-block (square), $(m+u)$-indexed, n-th order matrix. Its block-square domain $D^{\{M\times M\}}H_n$ is given by

$$ D^{\{M\times M\}}H_n = [D^{m\oplus u}H_n]_{m,u=1,\dots,M} \tag{A.11c} $$

Let us observe that the matrix (A.11) can be described in a generalized 'block-column' form. To do that, let us suppose that $D^{m\oplus u}H_n = D^mH_n \times D^uH_n$, where $(j_1,\dots,j_m) \in D^mH_n$ and $(k_1,\dots,k_u) \in D^uH_n$. Then we can write

$$ {}^{\{M\times M\}}H_n = [{}^{\{M\}\times u}H_n]_{u=1,\dots,M} \tag{A.12a} $$

where

$$ {}^{\{M\}\times u}H_n = \mathrm{col}\,[{}^{m\oplus u}H_n]_{m=1,\dots,M} \tag{A.12b} $$

or, equivalently,

$$ {}^{\{M\}\times u}H_n = [{}^{\{M\}}H_{n;k_1\dots k_u}]_{(k_1,\dots,k_u)\in D^uH_n} \tag{A.12c} $$

with

$$ {}^{\{M\}}H_{n;k_1\dots k_u} = [h_{j_1\dots j_m k_1\dots k_u}]_{(j_1,\dots,j_m) \in D^mH_n} \tag{A.12d} $$

<u>Equal matrices</u>

Two m-indexed matrices

$$ {}^mA = [a_{j_1\dots j_m}]_{(j_1,\dots,j_m) \in D^mA} \quad ; \quad {}^mB = [b_{j_1\dots j_m}]_{(j_1,\dots,j_m)\in D^mB} \tag{A.13} $$

will be called equal matrices if $D^mA = D^mB$, and if for each sequence of indices $(j_1,\dots,j_m) \in D^mA$

$$a_{j_1 \dots j_m} = b_{j_1 \dots j_m} \tag{A.14}$$

Sum of multi-indexed matrices

Given the matrices (A.13), we shall say that the m-indexed matrix

$$^mG = [g_{j_1 \dots j_m}]_{(j_1,\dots,j_m) \in D^mG} \tag{A.15a}$$

is the sum

$$^mG = {}^mA + {}^mB \tag{A.15b}$$

if $D^mG = D^mA = D^mB$, and if for each $(j_1,\dots,j_m) \in D^mA$

$$g_{j_1 \dots j_m} \stackrel{\Delta}{=} a_{j_1 \dots j_m} + b_{j_1 \dots j_m} \tag{A.15c}$$

Sum of block, multi-indexed matrices

Given two M-block (row), m-indexed matrices

$$^{\{M\}}A = [\,^mA\,]_{m=1,\dots,M} \quad ; \quad ^{\{M\}}B = [\,^mB\,]_{m=1,\dots,M} \tag{A.16a}$$

we shall say that the M-block (row), m-indexed matrix

$$^{\{M\}}G = [\,^mG\,]_{m=1,\dots,M} \tag{A.16b}$$

where ^mG is given by (A.15a), is the sum

$$^{\{M\}}G = {}^{\{M\}}A + {}^{\{M\}}B \tag{A.16d}$$

if for $m=1,\dots,M$ we have $D^mA = D^mB = D^mG$, and if for each sequence of indices $(j_1,\dots,j_m) \in D^mA$ the entries $g_{j_1\dots j_m}$ are expresed by (A.15c).

Product of a scalar and a m-indexed matrix

Given a scalar $c \in \mathbb{R}$ and a m-indexed matrix ^mA (A.13), we shall say that the m-indexed matrix ^mG (A.15a) is the product of the scalar and the m-indexed matrix

$$^mG = c \cdot {}^mA \tag{A.17a}$$

if for each $(j_1,\dots,j_m) \in D^mG$ (where $D^mG = D^mA$) we have

$$g_{j_1\dots j_m} \stackrel{\Delta}{=} c \cdot a_{j_1\dots j_m} \tag{A.17b}$$

Product of multi-indexed matrices

Let ^mA be given by (A.13), and let ^sB be given by (A.13) with m replaced by s, where $m \leq s$. Let υ, μ, ν be some given integers, satisfying

$$\upsilon + \mu = m \qquad ; \qquad \mu + \nu = s \tag{A.18a}$$

and moreover let $r = (\upsilon + \nu)$. Partitioning the m indices of the ^mA, and the s indices of the ^sB in accordance with (A.18a), and assuming that $D^mA = D^{\upsilon}A \times D^{\mu}A$ and $D^sB = D^{\mu}B \times D^{\nu}B$, we can write

$$^mA = [a_{k_1 \dots k_\upsilon j_1 \dots j_\mu}] \quad (k_1,\dots,k_\upsilon) \in D^{\upsilon}A \; ; \; (j_1,\dots,j_\mu) \in D^{\mu}A \tag{A.18b}$$

$$^sB = [b_{j_1 \dots j_\mu i_1 \dots i_\nu}] \quad (j_1,\dots,j_\mu) \in D^{\mu}B \; ; \; (i_1,\dots,i_\nu) \in D^{\nu}B \tag{A.18c}$$

We shall say that the $r = (\upsilon + \nu)$ -indexed matrix

$$^{r=(\upsilon+\nu)}G = [g_{k_1 \dots k_\upsilon i_1 \dots i_\nu}] \quad (k_1,\dots,k_\upsilon) \in D^{\upsilon}A \; ; \; (i_1,\dots,i_\nu) \in D^{\nu}B \tag{A.18d}$$

whose domain is $D^rG = D^{\upsilon}A \times D^{\nu}B$, is the μ-product of the matrices (A.18b,c)

$$^rG = {}^mA \cdot {}^sB = {}^{\upsilon \oplus \mu}A \cdot {}^{\mu \oplus \nu}B \tag{A.18e}$$

if $D^{\mu}A = D^{\mu}B = {}^{\mu}D$, and if for each $(k_1,\dots,k_\upsilon) \in D^{\upsilon}A$, and for each $(i_1,\dots,i_\nu) \in D^{\nu}B$

$$g_{k_1 \dots k_\upsilon i_1 \dots i_\nu} \triangleq \sum_{^{\mu}D} a_{k_1 \dots k_\upsilon j_1 \dots j_\mu} b_{j_1 \dots j_\mu i_1 \dots i_\nu} \tag{A.18f}$$

where the sum in (A.18f) denotes the μ-fold summation with respect to $(j_1,\dots,j_\mu)$ over the μ-variate index-set $^{\mu}D$.

Product of block, multi-indexed matrices

Given the M-block (row) matrix $^{\{M\}}A$ (A.9), and the $(M \times M)$-block square-matrix $^{\{M \times M\}}H$ (A.11), we shall say that the M-block (row), u-indexed matrix

$$ {}^{\{M\}}G = [{}^{u}G]_{u=1,\ldots,M} \quad ; \quad {}^{u}G = [\,g_{k_1,\ldots,k_u}]_{(k_1,\ldots,k_u)\in D^{u}G} \tag{A.19a} $$

(where for $u=1,\ldots,M$ we have $D^{u}G = D^{u}H$, with $D^{m\oplus u}H = D^{m}H \times D^{u}H$) is the block m-product of the matrices ${}^{\{M\}}A$ and ${}^{\{M\times M\}}H$

$$ {}^{\{M\}}G = {}^{\{M\}}A \cdot {}^{\{M\times M\}}H \tag{A.19b} $$

if for $u=1,\ldots,M$

$$ {}^{u}G \triangleq \sum_{m=1}^{M} {}^{m}A \cdot {}^{m\oplus u}H \tag{A.19c} $$

with $\cdot$ denoting the product (A.18e). Using (A.18) with $\upsilon=0$, $\mu=m$ and $\nu=u$, we can rewrite (A.19) as

$$ g_{k_1\ldots k_u} = \sum_{m=1}^{M} \sum_{{}^{m}D} a_{j_1\ldots j_m} h_{j_1\ldots j_m k_1\ldots k_u} \tag{A.19d} $$

where ${}^{m}D = D^{m}A = D^{m}H$ and $(k_1,\ldots,k_u)\in D^{u}G$. Equivalently, using (A.12d), we can write

$$ g_{k_1\ldots k_u} = {}^{\{M\}}A \cdot {}^{\{M\}}H_{k_1\ldots k_u} \tag{A.19e} $$

<u>'Outer' or Kronecker product of multi-indexed matrices</u>

Let ${}^{m}A$ be given by (A.13), and let ${}^{s}B$ be expressed by (A.13) with m replaced by s . We shall say that the $(m+s)$-indexed matrix

$$ {}^{m\oplus s}G = [g_{j_1\ldots j_m k_1\ldots k_s}]_{(j_1,\ldots,j_m)\in D^{m}A\ ;\ (k_1,\ldots,k_s)\in D^{s}B} \tag{A.20a} $$

(whose domain is $D^{m\oplus s}G = D^{m}A \times D^{s}B$) is the 'outer' (or Kronecker) product of the matrices ${}^{m}A$ and ${}^{s}B$

$$ {}^{m\oplus s}G = {}^{m}A \otimes {}^{s}B \qquad (A.20b) $$

if for each $(j_1,\dots,j_m) \in D^{m}A$ and $(k_1,\dots,k_s) \in D^{s}B$ we have

$$ g_{j_1\dots j_m k_1\dots k_s} \overset{\Delta}{=} a_{j_1\dots j_m}\, b_{k_1\dots k_s} \qquad (A.20c) $$

From (A.20) it follows that if

$$ {}^{1}Y = [y_j]_{\; j \in D^{1}Y} \qquad (A.21a) $$

then

$$ {}^{m}Y = \overset{m}{\otimes}\, {}^{1}Y = [\, y_{j_1} \dots y_{j_m} \,]_{(j_1,\dots,j_m) \in D^{m}Y} \qquad (A.21b) $$

where

$$ D^{m}Y = \underbrace{D^{1}Y \times \dots \times D^{1}Y}_{m} \qquad (A.21c) $$

<u>'Outer' product of block, multi-indexed matrices</u>

Let

$$ {}^{\{M\}}Y = [\,{}^{m}Y\,]_{m=1,\dots,M} \qquad (A.22a) $$

where ${}^{m}Y$ is given by (A.21), and moreover let

$$ {}^{\{M\times M\}}G = [\,{}^{m\oplus u}G\,]_{m,u=1,\dots,M} \qquad (A.22b) $$

with

$$ {}^{m\oplus u}G = [g_{j_1\cdots j_m k_1\cdots k_u}] \quad (j_1,\ldots,j_m) \in D^m Y \; ; \; (k_1,\ldots,k_u) \in D^u Y \tag{A.22c} $$

We shall say that the matrix ${}^{\{M\times M\}}G$ is the block, outer-product

$$ {}^{\{M\times M\}}G = {}^{\{M\}}Y \otimes {}^{\{M\}}Y \tag{A.22d} $$

if for $m,u=1,\ldots,M$

$$ {}^{m\oplus u}G \triangleq {}^{m}Y \otimes {}^{u}Y \tag{A.22e} $$

or, equivalently, if for each $(j_1,\ldots,j_m) \in D^m Y$ and $(k_1,\ldots,k_u) \in D^u Y$

$$ g_{j_1\cdots j_m k_1\cdots k_u} = y_{j_1}\cdots y_{j_m} y_{k_1}\cdots y_{k_u} \tag{A.22f} $$

APPENDIX 2

MULTI-VARIATE INDEX-SET RECURSIONS

LB-recursions: 'local' order-updates for the 'bi-variate' part of the L-forward index-set, and for the 'uni-variate' part of the B-backward index-sets; i.e., for $m=2,\dots,n+1$

$$L^{x}_{n,m} = L^{x}_{n,m-1} \cup L^{x+1,n-1}_{n,m-1} = \underbrace{\{x\} \cup \rlap{$\overbrace{\phantom{L^{x+1,n-1}_{n,m-2} \cup \{x+n+3-m,x+n+1\}}}^{L^{x+1,n-1}_{n,m-1}}$} L^{x+1,n-1}_{n,m-2}}_{L^{x}_{n,m-1}} \cup \{x+n+3-m,x+n+1\} \qquad \text{(A.23a)}$$

and for $m=n+2$

$$L^{x}_{n+1,o} = L^{x}_{n,n+1} \cup L^{x+1,n}_{n,n+1} = \underbrace{\{x\} \cup \rlap{$\overbrace{\phantom{L^{x+1,n-1}_{n,n} \cup \{x+1,x+n+1\}}}^{L^{x+1,n}_{n,n+1}}$} L^{x+1,n-1}_{n,n}}_{L^{x}_{n,n+1}} \cup \{x+1,x+n+1\} \qquad \text{(A.23b)}$$

These recursions can be schematically described as the LB index-set sections

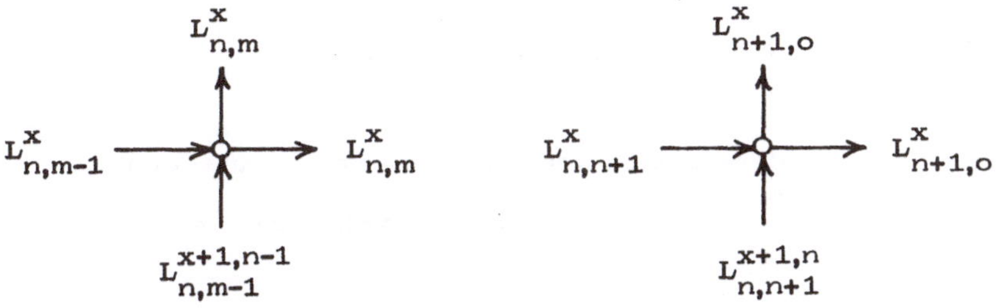

(A.23c)

BL-recursions: 'local' order-updates for the 'uni-variate' part of the B-forward index-sets, and for the 'bi-variate' part of the L-backward index-set; i.e., for $v=0$

$$L^{x,o}_{n,2} = L^{x,o}_{n,1} \cup L^{x}_{n,1} = \underbrace{\{x,x\} \cup \overbrace{L^{x}_{n,o}}}_{L^{x,o}_{n,1}}\overbrace{\cup \{x+n+1\}}^{L^{x}_{n,1}} \qquad \text{(A.24a)}$$

and for $v=1,\ldots,n$

$$L^{x,v}_{n,v+2} = L^{x,v}_{n,v+1} \cup L^{x,v-1}_{n,v+1} = \underbrace{\{x,x+v\} \cup \overbrace{L^{x,v-1}_{n,v}}}_{L^{x,v}_{n,v+1}}\overbrace{\cup \{x+n+1\}}^{L^{x,v-1}_{n,v+1}} \qquad \text{(A.24b)}$$

These recursions will be interpreted as the BL index-set sections

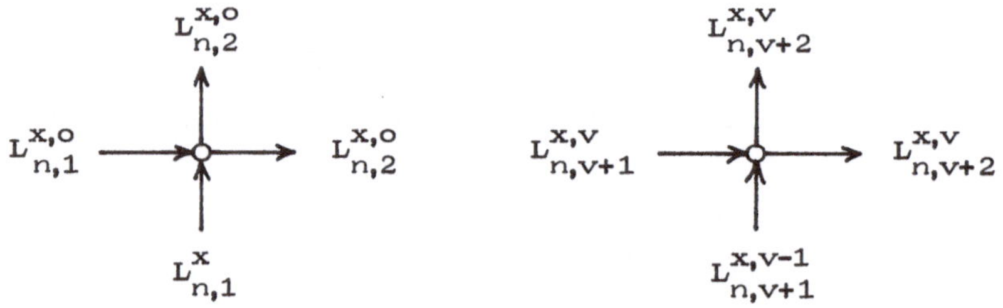

(A.24c)

BB-recursions: 'local' order-updates for the 'bi-variate' parts of the B-forward and B-backward index-sets; i.e., for $v=0$ and $m=3,\ldots,n+3$

$$L^{x,o}_{n,m} = L^{x,o}_{n,m-1} \cup L^{x}_{n,m-1} = \rlap{$\underbrace{\phantom{\{x,x\} \cup L^{x}_{n,m-2}}}_{L^{x,o}_{n,m-1}}$}\{x,x\} \cup \overbrace{L^{x}_{n,m-2} \cup \{x+n+4-m,x+n+1\}}^{L^{x}_{n,m-1}} \qquad \text{(A.25a)}$$

and for $v=1,\dots,n$ and $m=v+3,\dots,v+n+3$

$$L^{x,v}_{n,m} = L^{x,v}_{n,m-1} \cup L^{x,v-1}_{n,m-1} = \rlap{$\underbrace{\phantom{\{x,x+v\} \cup L^{x,v-1}_{n,m-2}}}_{L^{x,v}_{n,m-1}}$}\{x,x+v\} \cup \overbrace{L^{x,v-1}_{n,m-2} \cup \{x+n+4+v-m,x+n+1\}}^{L^{x,v-1}_{n,m-1}} \qquad \text{(A.25b)}$$

These recursions will result in the BB index-set sections

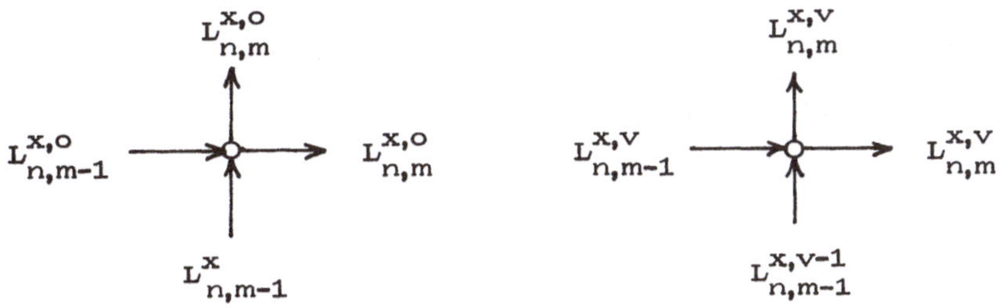

(A.25c)

'LOCAL' ORDER-UPDATE RECURSIONS

LB-recursions: for $m=2,\ldots,n+1$

$$\begin{bmatrix} A^{x}_{n,m}(Z) \\ \\ B^{x}_{n,m}(Z) \end{bmatrix} = \Theta^{x}_{n,m} \begin{bmatrix} A^{x}_{n,m-1}(Z) \\ \\ Z \cdot B^{x+1,n-1}_{n,m-1}(Z) \end{bmatrix} \qquad \text{(A.26a)}$$

$$\rho^{x}_{n,m} = (A^{x}_{n,m}(Z) \,,\, Z \cdot B^{x+1,n-1}_{n,m-1}(Z))_{\mathbb{Z}} \qquad \text{(A.26b)}$$

and for $m=n+2$

$$\begin{bmatrix} A^{x}_{n+1,o}(Z) \\ \\ B^{x}_{n+1,o}(Z) \end{bmatrix} = \Theta^{x}_{n+1,o} \begin{bmatrix} A^{x}_{n,n+1}(Z) \\ \\ Z \cdot B^{x+1,n}_{n,n+1}(Z) \end{bmatrix} \qquad \text{(A.26c)}$$

$$\rho^{x}_{n+1,o} = (A^{x}_{n,n+1}(Z) \,,\, Z \cdot B^{x+1,n}_{n,n+1}(Z))_{\mathbb{Z}} \qquad \text{(A.26d)}$$

BL-recursions: for $v=0$

$$\begin{bmatrix} A^{x,o}_{n,2}(Z) \\ \\ B^{x,o}_{n,2}(Z) \end{bmatrix} = \Theta^{x,o}_{n,2} \begin{bmatrix} A^{x,o}_{n,1}(Z) \\ \\ B^{x}_{n,1}(Z) \end{bmatrix} \qquad \text{(A.27a)}$$

$$\rho^{x,o}_{n,2} = (A^{x,o}_{n,1}(Z) \,,\, B^{x}_{n,1}(Z))_{\mathbb{Z}} \qquad \text{(A.27b)}$$

for $v=1,\ldots,n$

$$\begin{bmatrix} A^{x,v}_{n,v+2}(z) \\ \\ B^{x,v}_{n,v+2}(z) \end{bmatrix} = \Theta^{x,v}_{n,v+2} \begin{bmatrix} A^{x,v}_{n,v+1}(z) \\ \\ B^{x,v-1}_{n,v+1}(z) \end{bmatrix} \tag{A.27c}$$

$$\rho^{x,o}_{n,m} = (A^{x,o}_{n,v+1}(z)\ ,\ B^{x,v-1}_{n,v+1}(z))_{\mathbf{Z}} \tag{A.27d}$$

<u>BB-recursions:</u> for $v=0$ and $m=3,\ldots,n+3$

$$\begin{bmatrix} A^{x,o}_{n,m}(z) \\ \\ B^{x,o}_{n,m}(z) \end{bmatrix} = \Theta^{x,o}_{n,m} \begin{bmatrix} A^{x,o}_{n,m-1}(z) \\ \\ B^{x}_{n,m-1}(z) \end{bmatrix} \tag{A.28a}$$

$$\rho^{x,o}_{n,m} = (A^{x,o}_{n,m-1}(z)\ ,\ B^{x}_{n,m-1}(z))_{\mathbf{Z}} \tag{A.28b}$$

for $v=1,\ldots,n$ and $m=v+3,\ldots,v+n+3$

$$\begin{bmatrix} A^{x,v}_{n,m}(z) \\ \\ B^{x,v}_{n,m}(z) \end{bmatrix} = \Theta^{x,v}_{n,m} \begin{bmatrix} A^{x,v}_{n,m-1}(z) \\ \\ B^{x,v-1}_{n,m-1}(z) \end{bmatrix} \tag{A.28c}$$

$$\rho^{x,v}_{n,m} = (A^{x,v}_{n,m-1}(z)\ ,\ B^{x,v-1}_{n,m-1}(z))_{\mathbf{Z}} \tag{A.28d}$$

where

$$\Theta^{x}_{n,m} = (1-[\rho^{x}_{n,m}]^2)^{-\frac{1}{2}} \begin{bmatrix} 1 & -\rho^{x}_{n,m} \\ \\ -\rho^{x}_{n,m} & 1 \end{bmatrix} \tag{A.29a}$$

$$\Theta^{x,v}_{n,m} = (1-[\rho^{x,v}_{n,m}]^2)^{-\frac{1}{2}} \begin{bmatrix} 1 & -\rho^{x,v}_{n,m} \\ \\ -\rho^{x,v}_{n,m} & 1 \end{bmatrix} \tag{A.29b}$$

'LOCAL' FILTER SECTIONS

<u>LB-sections:</u>

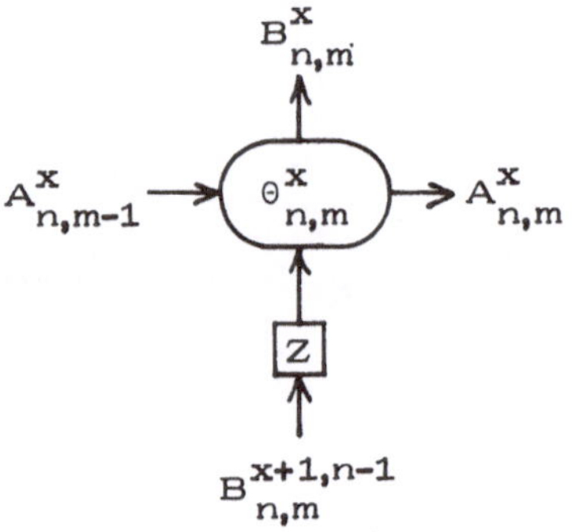

for m=2,...,n+1

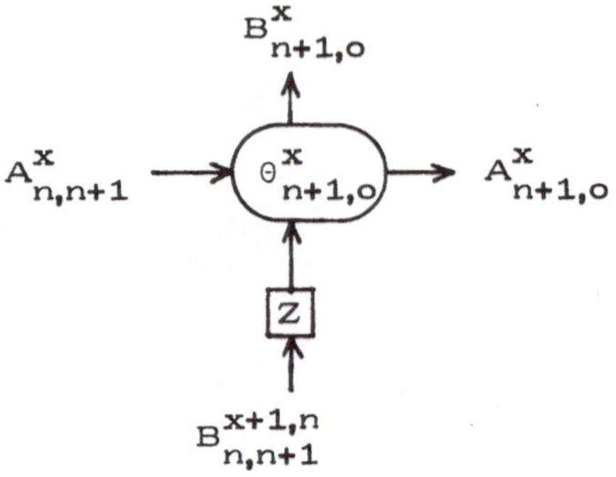

for m=n+2

<u>BL-sections:</u>

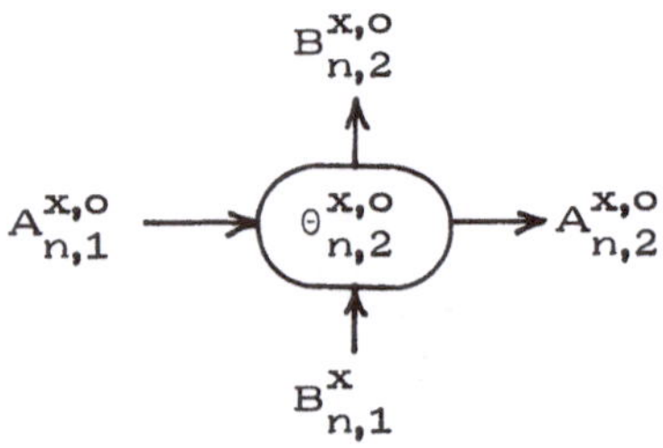

for v=0

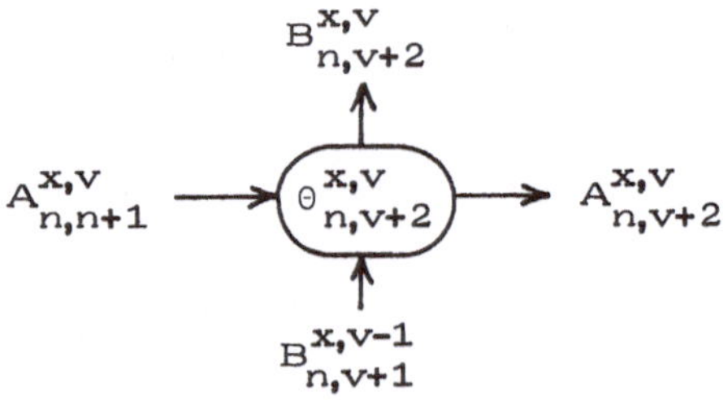

for v=1,...,n

<u>BB-sections:</u>

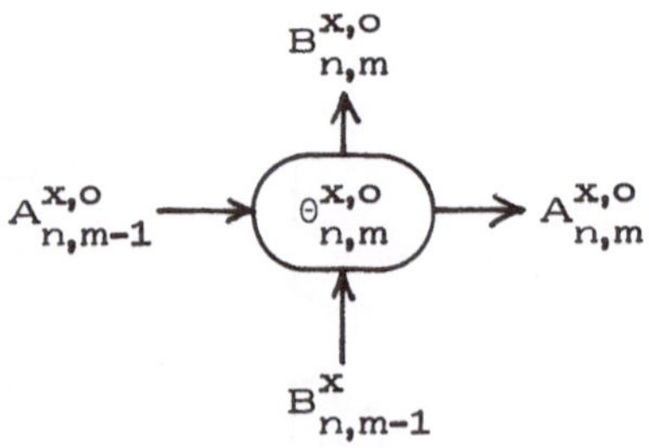

for v=0 and m=3,...,n+3

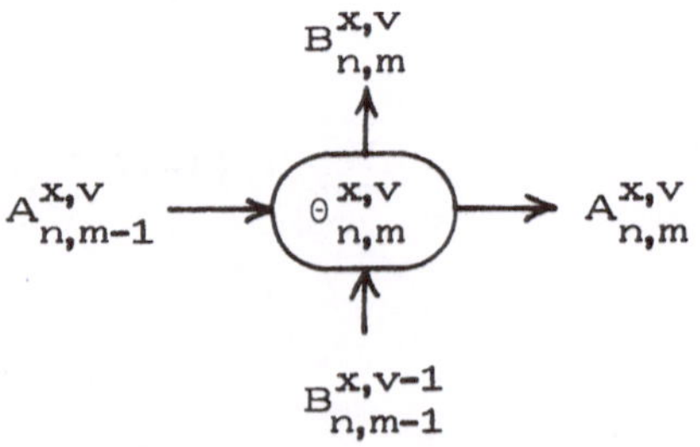

for v=1,...,n and m=v+3,...,v+n+3

Lecture Notes in Control and Information Sciences

Edited by M. Thoma

Vol. 43: Stochastic Differential Systems
Proceedings of the 2nd Bad Honnef Conference
of the SFB 72 of the DFG at the University of Bonn
June 28 – July 2, 1982
Edited by M. Kohlmann and N. Christopeit
XII, 377 pages. 1982.

Vol. 44: Analysis and Optimization of Systems
Proceedings of the Fifth International
Conference on Analysis and Optimization of Systems
Versailles. December 14–17, 1982
Edited by A. Bensoussan and J. L. Lions
XV, 987 pages, 1982

Vol. 45: M. Arató
Linear Stochastic Systems
with Constant Coefficients
A Statistical Approach
IX, 309 pages. 1982

Vol. 46: Time-Scale Modeling of Dynamic Networks
with Applications to Power Systems
Edited by J. H. Chow
X, 218 pages. 1982

Vol. 47: P. A. Ioannou, P. V. Kokotovic
Adaptive Systems with Reduced Models
V, 162 pages. 1983

Vol. 48: Yaakov Yavin
Feedback Strategies for Partially
Observable Stochastic Systems
VI, 233 pages, 1983

Vol. 49: Theory and Application of Random Fields
Proceedings of the IFIP-WG 7/1
Working Conference
held under the joint auspices of the
Indian Statistical Institute
Bangalore, India, January 1982
Edited by G. Kallianpur
VI. 290 pages. 1983

Vol. 50: M. Papageorgiou
Applications of Automatic Control Concepts
to Traffic Flow Modeling and Control
IX, 186 pages. 1983

Vol. 51: Z. Nahorski, H.F. Ravn, R.V.V. Vidal
Optimization of Discrete Time Systems
The Upper Boundary Approach
V, 137 pages 1983

Vol. 52: A. L. Dontchev
Perturbations, Approximations and Sensitivity Analysis
of Optimal Control Systems
IV, 158 pages. 1983

Vol. 53: Liu Chen Hui
General Decoupling Theory of Multivariable
Process Control Systems
XI, 474 pages. 1983

Vol. 54: Control Theory for Distributed
Parameter Systems and Applications
Edited by F. Kappel, K. Kunisch,
W. Schappacher
VII, 245 pages. 1983.

Vol. 55: Ganti Prasada Rao
Piecewise Constant Orthogonal Functions
and Their Application to Systems and Control
VII, 254 pages. 1983.

Vol. 56: Dines Chandra Saha, Ganti Prasada Rao
Identification of Continuous
Dynamical Systems
The Poisson Moment Functional
(PMF) Approach
IX, 158 pages. 1983.

Vol. 57: T. Söderström, P. G. Stoica
Instrumental Variable Methods
for System Identification
VII, 243 pages. 1983.

Vol. 58: Mathematical Theory of
Networks and Systems
Proceedings of the MTNS-83 International
Symposium
Beer Sheva, Israel, June 20–24, 1983
Edited by P. A. Fuhrmann
X, 906 pages. 1984

Vol. 59: System Modelling and Optimization
Proceedings of the 11th IFIP Conference
Copenhagen, Denmark, July 25-29, 1983
Edited by P. Thoft-Christensen
IX, 892 pages. 1984

Vol. 60: Modelling and Performance
Evaluation Methodology
Proceedings of the International Seminar
Paris, France, January 24–26, 1983
Edited by F. Bacelli and G. Fayolle
VII, 655 pages. 1984

Vol. 61: Filtering and Control of Random
Processes
Proceedings of the E.N.S.T.-C.N.E.T. Colloquium
Paris, France, February 23–24, 1983
Edited by H. Korezlioglu, G. Mazziotto, and
J. Szpirglas
V, 325 pages. 1984

Lecture Notes in Control and Information Sciences

Edited by M. Thoma

Vol. 62: Analysis and Optimization
of Systems
Proceedings of the Sixth International
Conference on Analysis and Optimization
of Systems
Nice, June 19–22, 1984
Edited by A. Bensoussan, J. L. Lions
XIX, 591 pages. 1984.

Vol. 63: Analysis and Optimization
of Systems
Proceedings of the Sixth International
Conference on Analysis and Optimization
of Systems
Nice, June 19–22, 1984
Edited by A. Bensoussan, J. L. Lions
XIX, 700 pages. 1984.

Vol. 64: Arunabha Bagchi
Stackelberg Differential Games
in Economic Models
VIII, 203 pages, 1984

Vol. 65: Yaakov Yavin
Numerical Studies
in Nonlinear Filtering
VIII, 273 pages, 1985.

Vol. 66: Systems and Optimization
Proceedings of the Twente Workshop
Enschede, The Netherlands, April 16–18, 1984
Edited by A. Bagchi, H. Th. Jongen
X, 206 pages, 1985.

Vol. 67: Real Time Control of Large Scale Systems
Proceedings of the First European Workshop
University of Patras, Greece, Juli 9–12, 1984
Edited by G. Schmidt, M. Singh, A. Titli,
S. Tzafestas
XI, 650 pages, 1985.

Vol. 68: T. Kaczorek
Two-Dimensional Linear Systems
IX, 397 pages, 1985.

Vol. 69: Stochastic Differential Systems –
Filtering and Control
Proceedings of the IFIP-WG 7/1 Working Conference
Marseille-Luminy, France, March 12-17, 1984
Edited by M. Metivier, E. Pardoux
X, 310 pages, 1985.

Vol. 70: Uncertainty and Control
Proceedings of a DFVLR International Colloquium
Bonn, Germany, March, 1985
Edited by J. Ackermann
IV, 236 pages, 1985.

Vol. 71: N. Baba
New Topics in Learning Automata
Theory and Applications
VII, 231 pages, 1985.

Vol. 72: A. Isidori
Nonlinear Control Systems:
An Introduction
VI, 297 pages, 1985.

Vol. 73: J. Zarzycki
Nonlinear Prediction
Ladder-Filters for Higher-Order
Stochastic Sequences
V, 132 pages, 1985.